DRIVEN TO EXTINCTION

DRIVEN TO EXTINCTION

The Impact of Climate Change on Biodiversity

RICHARD PEARSON

Published by the Natural History Museum, London

© 2011 by Richard Pearson
Originally published in the U.S. by Sterling Publishing Co., Inc. under
the title: DRIVEN TO EXTINCTION: The Impact of Climate Change
on Biodiversity

This edition has been published by arrangement with Sterling Publishing
Co., Inc., 387 Park Ave. South, New York, NY 10016

This edition published in 2011 by Natural History Museum, Cromwell
Road, London SW7 5BD

ISBN 978 0 565 09299 3

A catalogue record for this book is available from the British Library.

Design by Anderson Design Group AndersonDesignGroup.com
Printed by T J International Limited

Cover photo of a Panther Chameleon © Chris Mattison/age fotostock
Author photo by Denis Finnin © American Museum of Natural History

CONTENTS

PREFACE

This book is about the threat that climate change poses to the diversity of plants and animals that inhabit Earth. According to some headlines, more than a million species could face extinction due to climate change during the twenty-first century. Debate on this topic has tended to be polarized by the viewpoints of catastrophists who fret that we are heading toward total disaster, and skeptics who insist that there is nothing to worry about. My goal in this book is to describe as accurately as possible the current understanding of this issue. I aim to explain the science behind the debate and evaluate in an unbiased and level-headed manner the risks that climate change poses.

My main argument is that climate change is a severe threat to many species. In particular, climate change amplifies the risk of extinction when combined with other threats, such as habitat destruction, overharvesting, and invasive species. However, I show that it remains extremely difficult to predict how bad future impacts will be and I warn against alarmist messages of impending catastrophe. In doing so, I wade into some prickly—but important—territory concerning the role that scientists play in influencing public and political debate. Are scientists and environmentalists "crying wolf" over climate change?

There has been a huge amount of scientific research in recent years on the impacts of climate change, and my primary motivation for writing this book has been to communicate some of this fascinating and important science to a broader audience. The vast majority of information presented here is taken directly from articles published in peer reviewed scientific journals. Additional insights and tidbits of information that help make sense of the technical literature have been provided by colleagues (see Acknowledgments, page 223), but I have not adopted a journalistic approach of conducting interviews to provide content for the book.

It has not been my goal to provide an exhaustive review of studies that have been published on this topic. Rather, I have selected case studies that illustrate key concepts and that together link up to make a coherent, and I hope interesting, narrative. Wherever possible, I have selected studies to include a variety of organisms from different regions of the world, and I have paid particular attention to studies that illustrate how facts and figures provided by dedicated volunteers—"citizen scientists"—can be crucial for increasing our knowledge of the natural world.

In order to understand science, it is essential to understand scientific methods. I therefore aim to describe not only *what* we know, but also *how* we know it—the data, methods, and reasoning that led to a particular conclusion. For instance, I describe the various types of observations and models that have been used to predict future species' extinctions due to climate change. It is necessary to appreciate how these predictions are made if we are to interpret them correctly and make informed decisions about how to respond to the threat.

I hope that this book will appeal to anyone with an interest in nature conservation and/or the broader climate change issue. I find that the details of the science can be perfectly well communicated while avoiding jargon and an excess of technical terms. In fact, my experience in writing this book has taught me that, far from leading to imprecision, the translation of scientific articles into readable essays for the nonspecialist in some cases actually *increases* precision—there is no hiding flaky thinking or half-baked ideas behind technical terminology and impenetrable academic phrasing.

Climate change and nature conservation are—like it or not—issues that are important for all of us. We must be armed with a clear understanding of the risks involved if we, as a society, are to make informed and reasoned decisions about the action that should be taken to face the challenges of the coming century. My hope is that this book contributes toward deepening that understanding.

A note on units: Metric units, such as meters and kilometers, are most commonly used in science and I have therefore mostly adopted these measures throughout the book. However, in departure from the scientific literature, I present temperature measurements in degrees Fahrenheit rather than degrees Celsius. This is because Fahrenheit remains the most commonly used temperature scale in America, especially when referring to climate and weather. A list of conversions is included on page 223.

CHAPTER 1

A Climate for Life

Millions of species

There are approximately 1.5 million species living on our planet that are known to science. This is roughly the number of species that have been observed, named, and assigned a place on the tree of life by biologists. It's only a rough estimate because there is no centralized list, so we can't do a single giant tally. Nevertheless, it is reasonable to assume that we are correct to within a few hundred thousand.

The number of species that are recognized by science increases by about 10,000 each year. Again, we don't know the number for sure, but this is more or less the capacity of the world's biologists to name new specimens and write scientific descriptions about them. Contrary to what many people think, the discovery of unknown species is not at all rare. If you were to trek into a rarely visited tropical forest right now, you'd be pretty much guaranteed to find an organism that is unknown to science within a few hours. It's unlikely to be a bird or mammal, but finding an insect or plant is not unreasonable. If you care to sink to the bottom of the deep sea in a submersible, then I'll give you just a few minutes to find a new fish.

So, if there are still such riches to be discovered, what is the *actual* number of species on the planet? The sorry truth is that we have very little idea. A sensible estimate is probably around 10 million species, but an educated guess of 4 million is reasonable, while upwards of 100 million can be justified. A lot of the uncertainty stems from the fact that much of the planet is yet to be visited by teams of field biologists with their sampling nets and pickling jars. But the uncertainty is confounded by ongoing debate as to how species should be defined—when is one species really two similar species?[1]— and by ambiguity as to which types of organisms we should include in our tally—certainly mammals, birds, plants, and insects, but add to the mix microbes such as bacteria, and our estimate will be pushed up by a vast, but almost entirely unknown, amount. So, we can't be confident we're even in the right ballpark when estimating how many species there actually are—but there sure are many.[2]

These are a lot of big numbers to begin the book with, but bear with me: we're getting to know the stage on which the story will unfold. The myriad of species alive today represents the diversity of life on Earth, and it is this diversity—commonly referred to as our planet's "biodiversity"—that is of central interest in this book.

Another rough estimate is that only about one in a thousand of the species that have ever existed on Earth are still alive today.[3] We can't possibly know the proportion precisely, but the fossil record reveals that the vast majority of life forms that have ever been molded by evolution are long gone. The fossil record also reveals that many species went extinct during cataclysmic events scattered sporadically throughout Earth's history. Most famously, 65 million years ago something—most likely a massive meteorite smashing into the Earth—drove the dinosaurs to extinction, and there have been at least five other mass extinction events over the last 600 million years, each of which took a major chunk out of the diversity of species living at the time.

Here's my point: After each catastrophic event, it took evolution at least 10 million years to rebuild the former levels of diversity. That's a mighty long time, meaningless on human time frames. As far as we're concerned, the species alive today—however many millions there may be—are irreplaceable. Unless we put misguided faith in *Jurassic Park*–style technology to re-create extinct species from recovered strands of DNA, the current diversity of life is all we will ever have. It is the purpose of this book to assess the threat that climate change poses to this irreplaceable richness.

Climate clearly has a fundamentally important influence on nature, with species having adapted to live under certain conditions. For instance, emperor penguins have evolved a thick layer of blubber under their feathers to enable them to withstand some of the coldest temperatures on Earth, while their relatives, the Galápagos penguins, have been on an evolutionary diet to shed the insulation and survive in hot, equatorial climates. Likewise, you're not going to find palm trees on a skiing vacation or alpine conifers during a desert island getaway.

But these are extreme examples of how the climate affects species—differences between climates don't need to be so dramatic for us to see variation in species. Throughout nature, subtle differences in the climate have a major impact on the plants and animals that are found in a particular area.

Let's take a brief look at the flora and fauna of Vietnam as an example. Vietnam provides a good illustration because it is orientated north–south and therefore covers a range of latitudes, giving a gentle climate gradient across the length of the country. The north is relatively cool, with cold, humid winters and occasional frost on high ground. Summers are hot and wet, with humidity reaching levels that make you feel not just muggy, but downright soaked. Moving farther south, the temperature begins to warm and seasonal differences become less pronounced, so that by the time you reach the Mekong Delta in the far south, temperatures remain warm and stable throughout the year. This north-south climate gradient is also mirrored across elevation changes in the country, with the environment becoming predictably cooler and wetter as you move into mountainous regions, such as central Vietnam's Annamite Range.

These changes in climate are clearly reflected in the country's biodiversity. Northern regions contain forests of cold-tolerant plants, including birch and walnut trees, and spectacular rhododendron flowers. This environment is home to a number of animals that are *endemic* to the region; that is, they are found nowhere else on the planet. These endemics include the Red-Throated Squirrel, the Tonkin Snub-Nosed Monkey, and the White-Eared Night Heron—each named because of its peculiar physical characteristics, and each restricted to this cool environment. Moving south, the northern assemblage of species is gradually replaced by plants and animals favoring warmer and less seasonal climates. Southern Vietnam's biodiversity closely resembles that found in tropical regions, and includes a variety of species that are also often delightfully named to give us an insight into their most distinctive features: the Orange-Necked Partridge, the Grey-Faced Tit Babbler, and the Con Son Long-Tailed Macaque.[4]

Similar changes in species composition are evident along Vietnam's elevation gradients, with more cold-tolerant species found in upland areas. Of course, patterns in the distributions of species in Vietnam, and elsewhere, are not solely the result of the climate. Other factors, including fluctuations in sea level and the movement of continents over very long periods of time, have also played their part in scattering species around the landscape. But it is clear that the climate has a fundamental influence on where species are distributed.

What if the climate changes? What if environments in the south of Vietnam become too hot for species like the Orange-Necked Partridge and the Con Son Long-Tailed Macaque, and conditions farther north become more suitable for these species? Are many species destined to become extinct as their local environment becomes unsuitable, or will species relocate in order to find suitable conditions elsewhere? And what about cold-tolerant species that are isolated on mountaintops—will they be left with nowhere to go if the climate warms?

Before we begin to tackle these questions, I want to first emphasize why the issues raised are extremely important.

Trillions of dollars

These days, it is easy to feel disconnected from nature, to be unaware of and to forget the multitude of ways that we are dependent on biodiversity. Biodiversity is not a superfluous luxury—something to be enjoyed on weekends and holidays, or through TV wildlife documentaries. In reality, biodiversity provides services that are *essential* for human well-being. These services include the recycling of wastes, purification of drinking water, and maintenance of soil fertility. Biodiversity is also the source of food, fuels, building materials, and many medicines. Massive loss of species due to climate change would undoubtedly have profound implications for many of these services. If we imagine that biodiversity acts like a huge machine—a dynamic, interconnected system with many parts—then like any machine, it is inconceivable that the system will continue to function properly if we remove a substantial number of its parts.

One way to try to assess the importance of biodiversity is to assign an economic value to it. Those studies that have attempted to do so—by estimating the monetary worth of each of the many services provided and then totaling for the entire globe—have invariably come up with astronomically high numbers, measured in tens of trillions of U.S. dollars and comparable to the size of the entire global economy.[5] In truth, the economies of the world would grind to a halt without the services provided by biodiversity, so if climate change poses a threat to natural systems, then we should be more than a little concerned.

Of course, it is well known that our planet's biodiversity is under assault from a wide variety of human activities. Perhaps the most obvious and serious threat comes from the destruction of natural habitats. We have transformed at least half of the ice-free land surface of the planet for human use, turning natural forests and prairies into agricultural systems, and converting wetlands and even deserts into cities and towns.[6] The remaining isolated fragments of intact habitat cannot support the same diversity of life as large continuous expanses once did. Other major threats include the overharvesting of wild animals and plants for food, medicine, and building materials, and the (often accidental) introduction of non-native, invasive species in regions where they out-compete their local counterparts and drive endemic species to extinction.

A central point that I will emphasize in this book is that climate change should not be thought of as independent from other threats: it is the interaction of multiple threats that presents the biggest challenge to biodiversity. However, there are at least three key reasons why climate change presents unique challenges to conservation. First, climate change is a truly global phenomenon. Unlike other threats that may not affect a particular area, no region can expect to remain unaffected by the changing climate. For example, a nature preserve may be legally protected from habitat destruction and overexploitation, but climate change will not respect the preserve's boundaries. Second, we are already committed to climate change throughout the twenty-first century. The climate system takes many years to adjust to new levels of atmospheric greenhouse gases, so even if we halted emissions today, the quantity of greenhouse gases already in the atmosphere will continue to cause warming for decades to come. By contrast, if the political and social will existed, we could halt many other threats much more rapidly. Third, it is anticipated that the impacts of climate change on biodiversity will creep up on us gradually over the coming decades. The impacts will tend to be less sudden and immediately obvious than those caused by other factors, making climate change a kind of "stealth" threat. Society is likely to be slow to sense the danger if the changes seem gradual; instead, we tend to require a sudden jolt in order to provoke action to tackle an issue.[7]

Climate change catastrophe?

Climate change is one of the most challenging and controversial issues of our time. Much of the debate surrounding the issue revolves around the problem of uncertainty: How sure can we be that the climate is changing due to human activities? Are we certain that it will continue to get warmer? Most of the science we learn at school concerns neat, indisputable certainties, so it is sometimes difficult to appreciate that scientific frontiers are characterized by incomplete knowledge, debate, and by a healthy serving of uncertainty. To inform society about a pressing issue such as climate change, we must rely on weighing-up the available evidence and forming a consensus of opinions that best represents our current understanding.

In 1988, the United Nations responded to the need for a comprehensive and objective assessment of the climate change issue by establishing the Intergovernmental Panel on Climate Change (IPCC). The IPCC represents an immense collaboration between hundreds of experienced researchers from around the world, and although it has attracted controversy and criticism,[8] the panel remains the most authoritative source of information on climate change. The IPCC has now published four weighty (in fact, *very* weighty) assessment reports, the most recent in 2007 (a fifth installment is scheduled for 2013). These reports are crammed full of results from thousands of studies, along with carefully worded interpretations summarizing the consensus view. The authors express their assessment of the likelihood that a statement is correct using terms such as *very likely* (formally meaning a probability of more than 90 percent) or *likely* (probability of more than 66 percent).[9]

On the question of whether the climate is warming, the IPCC finds no need for fuzzy levels of confidence—the evidence is *unequivocal*. Observations of increases in global average air and ocean temperatures, widespread melting of snow and ice, and rising global sea level prove beyond reasonable doubt that the planet is heating up. However, it is important to note that this overall, planetary warming trend is reflected in varied regional changes; for instance, twentieth-century warming was strongest over the continental interiors of Asia and northern North America, while a few regions, such as an area over the northern North Atlantic (not too far from Greenland), have actually cooled. Similarly, there have been regional changes in rainfall patterns, with some areas getting wetter (including eastern parts of North and South America), while other areas have dried (including the Sahel region of northern Africa).

Identifying the cause of these changes is a little trickier. However, the IPCC has concluded that it is *very likely* that most of the observed increase in global average temperatures since the mid-twentieth century has been caused by human emissions of greenhouse gases. Atmospheric concentrations of carbon dioxide and other greenhouse gases have increased markedly as a result of burning fossil fuels—oil, coal, and gas—over the last two centuries. Indeed, data from Antarctic ice cores, which provide a record of ancient atmospheric constituents by preserving tiny bubbles of air trapped in the ice, reveal that greenhouse gas concentrations are now higher than they have been for at least the last 800,000 years. It is well established that increasing the atmospheric concentrations of greenhouse gases will lead to warming of the planet. This is because these gases only absorb radiation at certain wavelengths, meaning that more heat will reach the Earth's surface than will escape from it—the so-called *greenhouse effect*. The physical properties of greenhouse gases therefore point us toward the conclusion that human activity is having a discernible influence on the global climate.

As for what the future holds, the IPCC states that unless greenhouse gas emissions are reduced, it is *very likely* that changes in the global climate system will be larger during the twenty-first century than those observed during the twentieth century. Predictions of future climates are made for a range of alternative scenarios that take into account various uncertainties, including the quantity of greenhouse gases emitted by human activity over the coming years. Scenarios range from high emissions in a world of rapid economic growth driven by fossil fuel–intensive technologies, to low emissions in a more environmentally sustainable world that has adopted cleaner technologies. Predictions are therefore presented as a range of scenarios that try to put high and low bounds on what changes we can expect given a variety of plausible futures. These scenarios provide a basis for saying, "Well, this is the best we might expect, and that is the worst."

The IPCC estimates that under a best-case scenario, the rise in global average temperature during the twenty-first century is *likely* to be in the range of 2–5.2°F, and under a worst-case scenario, it is *likely* to be in the range of 4.3–11.5°F. To put these numbers in perspective, the average temperature each year in Washington, D.C., is around 2°F warmer than in New York City, while London is in the order of 11°F cooler than Barcelona. It is also predicted that the rise in temperature will be accompanied by other changes in the climate system. For instance: it is *very likely* that rainfall will increase at high latitudes and will decrease in most subtropical land regions; it is *very likely* that heat waves will continue to become more frequent; and it is *likely* that tropical storms (including hurricanes and typhoons) will become more intense.

These adjustments to the climate system are expected to impinge on many aspects of human life all around the world. To mention just a few of the potential impacts described by the IPCC: agricultural yields will decrease in warmer regions (including much of Africa) and increase in cooler regions; heat-related deaths will rise (largely due to cardiovascular mortality and respiratory illnesses caused by heat waves), and cold-related deaths will fall; the risk of wildfires will increase; shrinking of glaciers and snow cover will threaten freshwater supplies; some infectious diseases will move into new areas; and many millions of people will be displaced by flooding due to sea-level rise.

For the purposes of this book, I will accept the IPCC's conclusions that the climate is changing, that this is largely due to human emissions of greenhouse gases, and that the climate will continue to change throughout the coming century, having many far-reaching impacts. My goal is to assess in depth one element of these broader impacts: effects on biodiversity. On this topic, the IPCC's conclusion is that it is *likely* that 20–30 percent of species assessed so far will be at increased risk of extinction if global warming exceeds roughly 3.6°F, and that extinctions could be more on a scale of 40–70 percent if the temperature increase exceeds 6.3°F.[10]

In light of this conclusion, it appears that the outlook for biodiversity under climate change is extremely bleak. In general, the language used to talk about the impacts of climate change has become that of disaster and catastrophe. This was especially apparent in events preceding the 2009 UN Climate Change Conference in Copenhagen. At a United Nations summit convened in New York to gather momentum ahead of the conference, President Barack Obama said that by failing to meet the climate change challenge, "we risk consigning future generations to an irreversible catastrophe," while UN Secretary-General Ban Ki-moon similarly referred to "catastrophic climate change."[11] A few weeks later in London, UK Prime Minister Gordon Brown told the Major Economies Forum that the economic costs of failing to tackle climate change could be "greater than the losses caused by two world wars and the Great Depression."[12]

World leaders have consequently joined environmental pressure groups and the media in spreading a message of climate change doom and calamity. This message is especially apparent in some of the documentary-style feature films that have addressed the issue. Most famously, Al Gore's *An Inconvenient Truth* wraps sound science in a message of fear and crisis—the film's trailer declares that this is "by far, the most terrifying film you will ever see. . . . nothing is scarier . . ." And Gore is clear in his assessment of the magnitude of the problem: "Our ability to live is what is at stake." In another film—*The Age of Stupid*—Oscar-nominated actor Pete Postlethwaite plays an elderly man living in 2055 among the debris of a planet devastated by climate change. The film opens with images of the Sydney Opera House on fire and Las Vegas in ruins beneath desert sands. Postlethwaite's character has had the foresight to build a giant storage facility in the Arctic (no longer frozen, of course) in which he houses artifacts from the world's museums, including specimens from natural history collections that represent the past diversity of life on Earth. The image, then, is that of a Noah-like catastrophe. It seems we should be afraid, very afraid.

But from the perspective of biodiversity conservation, is the message of impending disaster alarmist or justified? What is the scientific reasoning behind the IPCC's conclusion that mass extinction is imminent, and is their conclusion reasonable?

Fingerprints of climate change

In building toward answering questions about future impacts, it is first necessary to understand how climate change is already affecting plants and animals around the world. In Chapters 2 through 4, I describe some of the known biological "fingerprints" of climate change—unique traits that provide compelling evidence that climate change has had an impact on biodiversity over recent decades.[13] In Chapter 5, I ask whether individual fingerprints from different regions of the world add up to a global picture that incriminates climate change as being a serious threat to biodiversity.

The first half of the book provides important groundwork for the second half, which looks more to the future. Chapter 6 gets to the heart of the matter by trying to quantify how many species are at risk of extinction due to climate change. I explain the methods commonly used to make future predictions and describe some of the alarming estimates that have been put forward. Then, in Chapters 7 and 8, I detail two key issues—potential for evolutionary adaptation, and indirect impacts within ecological communities—that could greatly affect the accuracy of predictions about future extinctions.

Chapter 9 addresses head-on the contention that scientists and environmentalists have exaggerated the seriousness of the issue, leading to alarmist headlines in the popular media. Finally, in Chapter 10, I focus on what can be done to manage the risk: what conservation strategies are likely to be effective for maintaining biodiversity over the twenty-first century?

Now, let's begin our search for fingerprints in Madagascar, with some of the most weird and wonderful creatures on the planet.

CHAPTER 2

Herps on Hills

Madagascar

The American Museum of Natural History occupies a large, rather grand building on the edge of Central Park in New York City. The museum celebrates the diversity of life on Earth with famous displays that include dinosaur bones, elephants, monkeys, lions, bears, birds, and a 28-meter-long blue whale. One of the museum halls is dedicated to amphibians and reptiles, which are often collectively termed *herptiles*—or *herps* for short. It is in this hall that you'll find a group of Komodo Dragons, a Galápagos Giant Tortoise, a 7-meter Reticulated Python, and, if you look carefully, a Panther Chameleon from Madagascar.

Most of the animals on display at the museum are real specimens, collected over many years during expeditions around the world. But the specimens that the public can view represent just a small fraction of what the museum houses in its collections: the herpetology department alone holds approximately 335,000 specimens, including individuals representing roughly 60 percent of all known herptile species. Cataloging biodiversity by building vast collections is an important role of natural history museums, so the museum in New York continues to organize expeditions to some of the planet's most biologically diverse regions. In some ways, these expeditions are the modern-day equivalent of the big-game hunting tours led by Teddy Roosevelt and others in the early 1900s, which brought back African mammals that are still on display at the museum. But while the old-school expeditions bring to mind images of pith helmets, safari suits, and hunting trophies, the ethos of modern trips is not only to catalog the diversity of life, but also to contribute to its conservation. As we'll see in this chapter, recent museum-led expeditions to catalog the herps of Madagascar have begun to shed new light on how climate change impacts biodiversity.

Amphibians and reptiles are a logical place to start searching for impacts of climate change because they are cold-blooded; that is, they are directly reliant on the environment for maintaining appropriate body temperature. If it's too cold, then they are unable to raise their body temperature enough to become active, and if it gets too hot, they will quickly feel the heat. Amphibians are also especially susceptible to drought since they are often reliant on ponds and streams—if there is no rain, then there are no ponds, and consequently no amphibians.

Madagascar, which lies off the southeast coast of Africa and is the world's fourth largest island, is home to many unique amphibians and reptiles. The island has been separated from other landmasses for a mighty long time—it was last connected to the African mainland around 120 million years ago, but, surprisingly, it was most recently connected to India, about 90 million years ago. This isolation has led to the evolution of a diverse array of life forms that are endemic to Madagascar. Indeed, Madagascar is recognized as one of the planet's *biodiversity hotspots* due to its unique and rich biodiversity, and also because this biodiversity is threatened by an expanding human population.[1]

The biodiversity hotspot label is well justified in terms of the island's herps, with more than 90 percent of its reptiles and 99 percent of its amphibians found only on Madagascar and its small offshore islands.[2] Perhaps the most famous reptilian inhabitants are the chameleons, which are locally revered for being mystic, no doubt due to their propensity to change color and their bizarre ability to move their eyes independently of each other. Another remarkable example of Madagascar's herps are the Leaf-Tailed Geckos, which have large triangular heads, big beady eyes, and a tail that bears an extraordinary resemblance to a leaf, providing one of nature's finest examples of camouflage.

Chris Raxworthy's fascination with amphibians and reptiles dates back to his childhood pets and now, as a curator of herpetology at the American Museum of Natural History, he is a world expert on these critters. Raxworthy is one of a number of field biologists who maintain the museum's long tradition of running expeditions in search of specimens to add to its collections. Since first visiting Madagascar as part of a student expedition in 1985, Raxworthy has returned to conduct herpetological surveys almost every year, visiting some of the most remote and mountainous areas of the island. When he first visited, only about 370 species of Malagasy herps had been identified, but over the ensuing years, Raxworthy and his colleagues have discovered more than 150 previously unknown species.

The identification of new species has been aided in recent years by powerful new laboratory techniques that enable biologists to look for genetic differences between specimens, but many of the techniques used in the field remain more primitive. How does an experienced herpetologist such as Raxworthy catch a fast-moving Leaf-Tailed Gecko? Answer: by stretching an elastic band between thumb and forefinger and flicking it at the unlucky creature. If your aim is good enough—and I dare say you need to be a pretty fine elastic-band flicker—the impact will stun the animal for long enough to catch it and take a sample of DNA or, sometimes, preserve the creature in a specimen jar.

The Tsaratanana mountain range in northern Madagascar includes the highest summit on the island—Maromokotro—which rises to 2,876 meters in elevation and is rich in species that are endemic to this region. Tsaratanana is particularly unusual in having an intact strip of natural habitat that extends from 1,400 meters in elevation to the summit of Maromokotro—a feature that is increasingly rare in a country where habitats are being destroyed at an alarming rate. This strip of habitat passes through a gradient of climates, with temperatures gradually dropping toward higher elevations. So if you set off up the trail that begins in Mangindrano village, you pass through a succession of different natural habitats, including rainforest and bamboo forest between 1,400 and 2,400 meters, grassy wetlands at around 2,500 meters, thick moss-laden forest around 2,600 meters, and rocky outcrops and heathland toward the summit. As you'd expect, the types of herps that you find also change as you hike up the hillside, with some species preferring warmer conditions at lower elevations, and other species hardy to cooler climes near the summit.

This strip of intact habitat at Tsaratanana has been surveyed by Raxworthy and his team on two separate occasions, first in 1993 and then again 10 years later.[3] The second survey followed exactly the same route as the first, with the researchers choosing identical locations to set up camp and search the surrounding vegetation. Armed with elastic bands, and headlamps for night searches, the survey team scoured all available habitats, paying particular attention to tree cavities, where herps tend to hang out. Some captured specimens were preserved in formalin (an aqueous solution of formaldehyde) and, most crucially, the exact location at which animals were found, including elevation, was recorded.

Back in the lab, the localities at which each species had been found were compared between the 1993 and 2003 surveys, and an interesting pattern stood out: it seemed that on the second visit, species tended to be found farther up the hillside than they had been a decade earlier. Close examination of the numbers confirmed that the midpoint elevation of each species' distribution—meaning the midpoint between the lowest and highest elevation recorded for each species—had increased by an average of 65 meters. Some species increased their elevation by much more than this (up to 400 meters) and a few species bucked the trend by moving downslope, but the dominant pattern over the 10 years was that of an uphill shift in distributions.

Shifts in species' distributions toward higher elevation are a fingerprint of climate change that we should logically expect: because temperatures gradually drop as you reach higher up a mountain, if the climate warms, then conditions currently found at low elevations will shift upslope to higher elevations. But is there evidence that the observed uphill shifts in Madagascar are the result of climate change?

We saw in the opening chapter that regional temperature trends do not always adhere to the global warming trend, so to answer this question we should look for evidence of warming in this part of the world. Unfortunately, there are not a great number of weather stations in Madagascar, and of those that exist, few have records spanning the years in which the herpetological surveys were undertaken. Still, with perseverance, Raxworthy and his colleagues were able to dredge up useful climate records from five Malagasy weather stations, all of which revealed a warming trend. By pooling these data with additional records from the broader Indian Ocean region, the research team concluded that northern Madagascar most likely warmed by somewhere between 0.18 and 0.67°F between the decades 1984–93 and 1994–2004.

This may not sound like much, but we can translate the observed warming into an expected uphill shift. If you take a thermometer on a hike up a hill you'll most likely record a regular drop in temperature of between 9 and 10.8°F for every 1,000 meters of elevation that you climb. This is known as the *lapse rate* and is caused by the decrease in air pressure as you get higher (as the pressure drops the air expands and there are fewer gas molecules to hold heat). A quick calculation tells us that the observed temperature change in Madagascar (0.18–0.67°F) equates to an uphill shift of 17–62 meters if we assume a lapse rate of 10.8°F per 1,000 meters, or 20–74 meters if we assume a lapse rate of 9°F. This, of course, is uncannily close to the average uphill displacement of 65 meters observed for the herps at Tsaratanana.

So it seems to add up that the observed distribution shifts are due to climate change. Raxworthy and his colleagues searched for possible alternative explanations, yet climate change remained the most likely culprit. Thus, we can expect the upslope trend to continue into the future as warming progresses. In later chapters, we'll look at some neat ways that fairly complex computer models can be used to predict future impacts, but for now we can forecast what is likely to happen over the coming decades based on the simple lapse rate.

Taking a temperature rise of 2°F—which is a fairly middle-of-the-road projection for the twenty-first century, according to the IPCC[4]—and applying the lapse rate of 9–10.8°F per 1,000 meters, we can expect to see upslope distribution shifts in the order of 300 meters over the coming century. That's rather a lot in comparison to the observed shifts from the recent surveys, and it doesn't bode well for the survival of upland species. Throughout Madagascar, many endemic species are found only around the highest summits, a trend that is true not only for the herps, but also for numerous other groups including plants, freshwater insects, and rodents. Upslope distribution shifts aren't an option for these species, thus bringing into play a high risk of extinction. At Tsaratanana, for example, two species of frog and one species of gecko are endemic to sites within 300 meters of the summit. A temperature rise of 3°F would push conditions that are suitable for these species up and off the top of the mountain.

Could many of Madagascar's unique herps, including the chameleons and the Leaf-Tailed Geckos, be heading toward extinction in this way? Interestingly—perhaps prophetically—Raxworthy and his team did not find two of Tsaratanana's three high-elevation endemics during their 2003 surveys. It's too early to consider them extinct—the team might have failed to find them even if they were still there—but the signs are not encouraging.

In order to find evidence that species have been driven to extinction by climate change, we can shift our focus to the herps of Central and South America.

The lifting cloud base

Monteverde encompasses an area of tropical montane cloud forest in the Tilarán Mountains of Costa Rica. The forests are spectacular. Trees are tightly packed and tend to have gnarled trunks, dense crowns, and small, thick leaves. Mosses, lichens, and vines are abundant, blanketing tree trunks and dangling free in the moist air. Needless to say, biodiversity and endemism are sky high.

The Monteverde area was settled in 1951 by a group of about 50 North American Quakers, some of whom had a strong conservation ethic. In 1972, through efforts of the Quaker community and international donors, the first preserve was founded in the area—the Monteverde Cloud Forest Preserve—affording this unique forest ecosystem some legal protection. The preserve covers an area of roughly 40 square kilometers and almost the entire natural habitat remains intact. Against a backdrop of widespread habitat destruction, this should be a safe haven for the local flora and fauna.[5]

But in the late 1980s, biologists working in the area—especially the herpetologists—began noticing that things were going awry. Species' populations throughout Monteverde were in rapid decline. In particular, a sudden crash of populations in 1987 affected numerous species, many of which never bounced back and now seem to be lost from the forests. The disappearance that has drawn the most attention is that of the Golden Toad. This brightly colored species was endemic to the high ridgetops at Monteverde, so its vanishing here would represent a global extinction—total, irreversible loss of a species.

The Golden Toad wasn't even known to science until the mid-1960s, when University of Miami biologist Jay Savage coined the scientific name *Bufo periglenes*, which translates from its Greek origins literally as "bright toad." On first encountering the toad in 1964, Savage was so astonished by the coloration that he remarked how it looked as if someone had dipped each one in enamel paint. And they were abundant, too—everywhere he looked he saw bright blotches of orange standing out against the black soil.[6] In the years following Savage's first visit, populations of the Golden Toad were monitored and remained healthy. But they never recovered from the 1987 crash. The last sighting was in 1989 and although there remains a glimmer of hope that a few individuals may remain hidden in the forest, the species is now presumed to be extinct.

The Golden Toad lends its name to the Golden Toad Laboratory for Conservation, a research and educational facility located just a few hundred meters from the entrance to the Monteverde Preserve. Facilities at the lab include a greenhouse with a central pond, a stream with waterfalls, and a sprinkler system for producing artificial rain. A sound system can even play rainforest sounds, such as frog calls and thunderstorms, to re-create the natural environment as closely as possible. The facility will soon undergo remodeling,[7] but in the past, the greenhouse has been used to breed frogs and provide a stunning educational exhibit. The key thing about this human-made microcosm is that the climate can be carefully controlled—the temperature responds to the turning of a thermostat and the rains come at the push of a button. Accidentally getting a setting wrong would quickly result in a greenhouse full of dead frogs.

Nobody understands this fragile relationship between climate and amphibian survival better than resident scientist Alan Pounds. Years of experience studying Monteverde's herps led Pounds and his colleagues at the lab to suspect that climate change may have had a hand to play in the disappearance of the Golden Toad. It took some years to amass supporting data, but by 1999 the researchers had crafted a strong argument supporting their suspicions.[8]

The argument revolves around a theory that Pounds terms the "lifting-cloud-base hypothesis." The chain of thought behind this hypothesis is as follows: The Monteverde Cloud Forest receives its abundant moisture from water vapor blown inland off the Caribbean Sea. Once the moisture reaches the highlands, it condenses, forming the clouds that give the cloud forest its name and releasing heat that warms the atmosphere. As ocean temperatures have increased in recent decades, evaporation from the warm ocean surface also has increased, resulting in larger amounts of water vapor and, hence, heat, reaching the cloud forest. The expected result is atmospheric warming that is more pronounced than in other regions of the globe, meaning that Monteverde should be experiencing at least its fair share of global climate change.

Now, as we've already seen in Madagascar, warming in a mountainous region causes temperatures previously found at lower elevations to move uphill. Since the formation of clouds is linked with air temperature, a related consequence of warming is that the height at which clouds form will also move uphill. In Monteverde, clouds form as the trade winds flow upward and cool on meeting the slopes of the Cordillera de Tilarán, so it is reasonable to expect that these winds are now flowing higher up the cordillera before reaching the temperature at which condensation kicks in. The result should be a lifting of the forest's clouds (hence the "lifting-cloud-base hypothesis") and since mist from clouds provides life support for moisture-dependent amphibians, this could explain the disappearance of the Golden Toad.

That's the theory, so Pounds and his lab colleagues set

about finding and analyzing data that might support, or contradict, their hypothesis.

Here Pounds had some luck. One of the original Quaker settlers in Monteverde, John Campbell, had recorded daily rainfall at a site near the entrance to the forest preserve since the early 1970s. Such a set of records is like gold dust to a biologist interested in climate change. Because days without rainfall represent days without mist, the observed pattern of rainfall conveys information about the frequency of clouds in the forest.

Careful analysis of these records for the dry season (which runs from January to April) uncovered a clear trend exactly in keeping with the lifting-cloud-base hypothesis: it turned out that the number of days when the forest was doused in clouds had gradually decreased over the previous 25 years. Perhaps most importantly, the records also showed that days without mist had increasingly occurred one after the other, with back-to-back mist-free days resulting in dry periods lasting up to sixteen days. That is an awfully long time for a frog to go without moisture. Researchers at the Golden Toad Lab would never have dreamed of leaving the rains turned off in their greenhouse for so long.

So there was strong evidence in support of the lifting-cloud-base hypothesis. Climate change is therefore implicated as a key factor behind the amphibian population crashes of 1987, and the Golden Toad has now achieved unfortunate iconic status as the first documented extinction of a species due largely to recent climate change.

Publication of these findings from Monteverde sent shock waves around the conservation community. Amphibians are a particular conservation concern, with many reports of declining populations from around the world. A major worldwide analysis known as The Global Amphibian Assessment, published in 2004, identified 427 species as being "critically endangered," including 122 species that are "possibly extinct."[9] The thing that is particularly worrisome about the plight of amphibians is that many populations have declined even in areas where there has not been habitat destruction. This was the case at Monteverde, leading many researchers in other parts of the world to go back and reassess their data in light of the new hypothesis. Could climate change be a major contributor to amphibian declines worldwide?

One important stumbling block that soon emerged was that amphibian declines were not exclusively occurring in high elevation cloud forest environments. In fact, many declines were being recorded at lower elevations, among populations that were not reliant on mist for survival. The lifting-cloud-base hypothesis did not therefore provide a general explanation for extinctions in different regions and at different elevations. It seemed that things were a little more complicated, and it took Pounds another 7 years, until 2006, before he had the data to support a broader, more comprehensive hypothesis, this time labeled the "climate-linked epidemic hypothesis."[10]

A climate for disease

In order to build a detailed picture of amphibian declines that extended beyond Monteverde, Pounds teamed up with a large group of herpetologists who have tracked population changes in different areas throughout Central and South America. The researchers decided to focus on one particular group of amphibians, the Harlequin frogs,[11] which are distributed from Costa Rica south as far as Bolivia, and east into the Amazon Basin. The species in this group are easy to find and identify due to their bright, distinctive coloring (which is often a warning sign that a species is noxious), and by all accounts, they are also slow to escape from prowling herpetologists. So data available for these species was particularly abundant.

In total, 75 researchers pooled their data, providing unprecedented detail regarding the status of all 113 known species of Harlequin frogs, some of which were still awaiting formal description and naming in the scientific literature. The records indicate that around 30 of the species have been missing from all known localities for at least eight years, long enough to fear that these species may be extinct. On top of this, population sizes for another 42 species have been reduced by more than half in recent decades. Of the remaining 40 or so species, only 10 are known to have stable populations (the data isn't sufficient to judge the status of the others). All in all, it's fair to say that the Harlequin frogs are in crisis.[12]

Although the lifting-cloud-base hypothesis does not provide a general explanation for the crisis, Pounds still suspected that climate change might be an important factor because the data showed that species tended to disappear soon after a relatively warm year. Take Ecuador's Jambato Toad (a species of Harlequin frog), for example. Much like the Golden Toad, this species used to be fairly abundant but was last seen in 1988, following the uncommonly hot and dry conditions of 1987.[13] Overall, around 80 percent of Harlequin frog species that are feared extinct were last seen immediately after a relatively warm year. Pounds didn't discover a perfect, straightforward association—populations didn't *always* crash when a warm year came along—but there certainly seemed to be a link between high temperatures and the last known sighting. In fact, Pounds and his colleagues calculated that the likelihood of the observed association occurring by chance is less than one in a thousand.

So climate change was definitely implicated, but how exactly are the Harlequin frogs being affected? Here there seemed to be two things that needed explaining. First, it was odd that there wasn't a more direct link between population crashes and temperature—why should populations get through some warm years unscathed and then crash another year when the weather wasn't a great deal worse? Second, analysis of the elevations at which populations were going extinct revealed an unexpected pattern: the proportion of populations dying out was highest at mid-elevations. In fact, upwards of 90 percent of species previously found at mid-elevations had been lost, while in the lowlands and highlands this figure was closer to 60 percent. Why should extinctions be most severe at mid-elevations?

Let's begin by addressing the first of these questions, concerning the lack of a direct link with temperature. Pounds and his colleagues came up with an imaginative solution by proposing that climate change might increase the likelihood of disease epidemics, thus indirectly impacting frog populations. Specifically, they suggested that climate change might promote outbreaks of a fungus called *Batrachochytrium dendrobatidis*, which is the cause of an infectious disease—chytridiomycosis —that causes mortality in amphibians worldwide.

Frogs are thought to contract chytridiomycosis when they come into contact with infected water, where the fungus is able to invade the surface layers of their skin. Frogs drink through their skin, and so the fungus reduces their ability to absorb electrolytes, eventually causing heart failure.[14] Pounds and his colleagues were well aware that chytridiomycosis is common throughout the range of the Harlequin frogs, and bringing the disease into the equation might explain the lack of a one-to-one association between climate and population crashes: a Harlequin population might survive a warm year if the fungus is absent, but the chance of being hit by an outbreak of chytridiomycosis may be much higher when conditions are unusually warm.

Now, what about the mid-elevation conundrum? It turns out that chytridiomycosis provides a good explanation for that, too. Digging around in the literature, Pounds and his colleagues found evidence from lab experiments that the disease-causing fungus grows best between 63 and 77°F, peaking at around 73°F and dying at 86°F. Comparing these optimum conditions against temperature records from localities spread throughout the range of the Harlequin frogs, it was apparent that maximum daily temperatures commonly exceed the 86°F threshold in the lowlands, while minimum night-time temperatures often drop too low for the fungus in the highlands. So, we have reason to believe that Harlequin frog populations are more likely to suffer from chytridiomycosis outbreaks at mid-elevations, where temperatures are neither too hot nor too cold.

If we next take a look at how minimum and maximum temperatures have changed in the region, we find that there has been a reduction in daily temperature extremes since the mid-1970s: days have become cooler and nights have become warmer. The likely reason for this once again concerns the affect of climate change on clouds; however, the key factor this time is not the height at which clouds form (as was the case at Monteverde), but rather the influence of increased cloudiness on daily temperature cycles. Rising temperatures increase the amount of water in the atmosphere (due to increased ocean surface evaporation and the higher capacity of air to hold water vapor) which produces more clouds. This, in turn, affects daily temperature extremes, since clouds keep the day cool by blocking solar radiation and the night warm by reducing heat loss, much like a blanket. The expected outcome of all this is a reduction in temperature extremes, which matches the observed record.

The overall conclusion, then, is that climate change is causing temperatures to converge on a range that is just right to promote disease outbreaks: the growth of *Batrachochytrium dendrobatidis* is favored at mid-elevations, resulting in more chytridiomycosis epidemics and, ultimately, a crisis among Harlequin frog populations.

Loading the dice

The climate-linked epidemic hypothesis provides an elegant explanation for the crisis among Harlequin frogs, but it is by no means the full story and remains the subject of ongoing debate.[15] Alan Pounds recognizes that the focus on a single disease—chytridiomycosis—is an oversimplification. In truth, a number of studies from around the world have documented frog population crashes without finding any evidence of a chytridiomycosis outbreak. What's more, frog declines have often been accompanied by equally mysterious declines in lizard populations, and it is unlikely that the same fungus that attacks moist-skinned amphibians is responsible for killing dry-skinned reptiles.[16] So the link between climate change and chytridiomycosis outbreaks certainly does not provide a general explanation for declines in amphibians and reptiles worldwide.

We don't even need to leave Costa Rica to find yet another possible explanation for herptile declines; instead, we can turn our attention away from the high and middle elevations, toward the lowlands. At La Selva forest reserve in the Caribbean lowlands of Costa Rica, records dating back to the 1950s reveal that the number of amphibians and reptiles has decreased by roughly three-quarters in recent decades. As at Monteverde, this is a protected forest, so habitat destruction can't explain the dramatic declines. Nor is there any evidence that chytridiomycosis has infected these populations. Instead, it seems likely that the decline in herps is related to a reduction in the amount of leaf litter, the layer of decaying leaves that carpets the forest floor. Many species of amphibians and reptiles live exclusively among the leaf litter, so a reduction in the amount of decaying matter will have a negative impact on these populations.

And what could cause a reduction in leaf litter? Well, for a start, things generally decompose faster in warmer and wetter conditions, so increases in temperature and rainfall will cause leaves that have fallen to disappear more quickly. And next, leaf fall in tropical forests tends to be stimulated by drought, so more rainfall throughout the year will result in fewer leaves from which the leaf litter can form. We've already heard plenty about changes in cloud cover, temperature, and precipitation in Costa Rica, but as a final note, records for this lowland site confirm that conditions have become consistently warmer and wetter since the early 1970s. So it seems likely that climate change is causing declines in these lowland herp populations, this time by reducing the amount of leaf litter, on which many species are dependent.[17]

We've now looked at a number of potential threats to amphibians and reptiles, including upslope distribution shifts, changes in mist frequency due to a lifting cloud base, disease epidemics, and reductions in the quantity of leaf litter. Although no single factor provides an explanation for population declines worldwide, or across all elevations, all of the threats I've discussed share the common trait of a link with climate change. There are undoubtedly other key threats that are not linked with climate change, most notably the destruction of habitats, but the evidence we have to date strongly suggests that climate change is playing an important role in the current crisis among amphibians and reptiles. It seems there are multiple mechanisms by which climate change impacts populations, and these remain only partially unraveled by science. But the broad picture of climate change as an underlying threat that puts extra pressure on species is becoming clear. In the words of Alan Pounds, climate change "loads the dice" against species' survival.[18]

The complexities of this game of chance make it difficult to predict exactly how climate change will impact herps—and biodiversity, more generally—in the future. For instance, nobody would have predicted the following ironic chain of events: Some of the earliest records of chytridiomycosis are from African Clawed Frogs in South Africa in the 1930s. These frogs became a global commodity during the 1940s and 1950s because it was discovered that they could be effectively used for human pregnancy tests. (A sample of the woman's urine is injected under the frog's skin; if the woman is pregnant, a hormone in her urine will cause the frog to lay eggs within a few hours.) Analysis of old specimens stored in natural history museum collections has shown that the time when African Clawed Frogs were being shipped around the world was also the time when chytridiomycosis achieved a global distribution. So it seems that the expansion of one species—in this case due to human ingenuity—may have spread a disease that ultimately led to the extinction of numerous other species.[19] The potential for unexpected, indirect consequences such as this to arise due to climate change will reappear as a common theme in later chapters.

It seems fitting to close this chapter with the case of the African Clawed Frog, in which data from natural history museum collections were crucial for unraveling the story. We began the chapter in the halls of the American Museum of Natural History, where vast collections of specimens continue to be accumulated in order to catalog the diversity of life on Earth. In light of the evidence we've now seen of a crisis among amphibian and reptile populations, and the connection with climate change, it seems reasonable to question how much of herptile diversity is likely to be wiped out over the coming decades. Are Madagascar's endemic chameleons and Leaf-Tailed Geckos heading toward the same fate as Monteverde's Golden Toad? Perhaps specimens preserved in jars at the American Museum's herpetology department will be our only record of many of these species by the end of the century.

CHAPTER 3

To the Ends of the Earth

A long-term perspective

Geological time scales are mind-boggling. Consider that multicellular organisms first appeared in the fossil record around 700 million years ago, the first plants spread across the land surface around 420 million years ago, and vertebrate animals first crawled onto land around 350 million years ago. Still, in geological terms, this is all relatively recent; for instance, the 65 million years that have passed since the demise of the dinosaurs represent just 1.4 percent of the age of the Earth.[1]

Periods of such magnitudes help us to put modern-day climate change into perspective. Life on Earth has endured many huge swings in global climates over millions of years, and it is tempting to discount recent changes as being insignificant in the grand scale of things. I'll return to this point toward the end of the book, where we'll look at ways in which contemporary climate change differs from climate shifts in the distant past. (In particular, we'll see how climate change today combines with other factors, such as habitat destruction and overfishing, to threaten biodiversity.) But for now, our interest is in understanding *how* species responded to climate change in the past—what does a long-term perspective tell us about how species respond to a changing climate?

In geological terminology, the *Pleistocene* refers to the period stretching from roughly 2 million years ago until 11,500 years ago, which is characterized by the repeated advance and retreat of glacial conditions. Ice sheets of the Pleistocene reached over large parts of Europe, Asia, and North America, and global mean temperature swung by as much as 11°F between the warm interglacial periods and the frigid glaciations. Not surprisingly, these climatic swings had a huge impact on the planet's flora and fauna. During interglacial stages, vegetation was somewhat similar to today. During glacial stages, the northern continents were dominated by low-lying herbaceous vegetation, similar to that which is restricted to arctic regions today.

We have an especially detailed understanding of how species responded to climate change during the warming that took us out of the last glaciation, around 11,500 years ago. This is because the fossil record for the last 20,000 years or so is remarkably complete, especially for plants. And we have one of the smallest possible fossils to thank for this: pollen.

Many plants produce huge quantities of pollen, which they spread liberally around the area with a little help from the wind, and often from animals including birds and insects. Some pollen inevitably falls into lakes or peatlands, where it accumulates alongside other sediments, settling layer upon layer, year after year. Because lake and peat sediments remain wet, they lack oxygen and so they are unfavorable for the organisms that cause decomposition. Because of this, pollen can remain fossilized for many thousands of years, making it possible to reconstruct vegetation changes by extracting sediment cores and studying the different kinds of pollen that are preserved at different depths.

In fact, fossil pollen can also be retrieved from dry environments, this time with thanks to the humble packrat. Packrats are small rodents common in deserts and highlands throughout North America. They build their nests (known as "middens") from virtually anything they can get their paws on, including leaves, fruits, twigs, and other debris with which pollen inevitably gets mixed. Because of the dry conditions they inhabit, packrats produce very viscous urine, which they spread liberally around their nest. When this urine dries, it crystallizes and cements the nest together, preserving the nest's building materials—including pollen—in a time capsule, much like amber preserves insects that became stuck in ancient tree sap.

So nature has left us with a fairly good picture of how vegetation responded to warming at the end of the last glacial stage. One of the responses seems to have been extinction. Among animals, the loss of large mammals such as the woolly mammoth and the saber-toothed tiger is well documented, but it is difficult to discern whether these extinctions were mostly the result of climate change or of overkill by expanding human populations. More informative for our purposes is the evidence of the extinction of plants, since the disappearance of certain plant species from the fossil record is less easy to ascribe to direct exploitation by humans.

A good example of a plant that went extinct during the last deglaciation is a species of spruce—*Picea critchfieldii*. This species used to be widespread across eastern North America, but it entirely disappears from the fossil record at the end of the last glacial stage. Spruces are large evergreen conifers that are characteristic of cool climates—roughly 35 spruce species are still in existence and they are distributed throughout regions such as the Alps, the Himalayas, and Scandinavia—so it appears that *Picea critchfieldii* was a casualty of climate warming that made much of its range uninhabitable.[2]

A long-term perspective therefore demonstrates the potential for extinction due to climate change, but this is not the most prevalent trend seen in the fossil pollen record. In fact, the majority of plant species responded to climate change by undergoing extensive range changes. And I mean *extensive* range changes. The pollen record shows that many tree species shifted their ranges by hundreds, or even thousands, of kilometers between the last glacial period and the current interglacial. Range shifts in plants are achieved through seed dispersal into areas that have become climatically suitable, leading to the establishment of new populations, while populations in regions that become unsuitable eventually die out. The result, over the course of a number of generations, is a shift in the species' distribution.

Large shifts in distribution during the transition to current climate conditions have been mapped for many plant species. For example, species of spruce in North America were found mostly in the central United States around 20,000 years ago, but by 7,000 years ago those species that were able to keep pace with the changing conditions—the luckier relatives of *Picea critchfieldii*—were restricted to more northerly reaches of Canada, where they are common today.[3] Similarly in Europe, oak trees were restricted to southern Spain and Italy during the last glacial period but have expanded as far north as

Norway and Sweden under warmer present-day conditions.[4] And it seems that trees were not alone in shifting their ranges —other taxonomic groups for which there is a good fossil record, including terrestrial mammals, mollusks, and beetles, show similarly extensive shifts.[5]

These examples demonstrate another fingerprint of climate change: as the climate warms, species shift their ranges toward higher latitudes—that is, toward the poles. Local patterns may be more complicated, but, in general, species track suitable conditions northward in the northern hemisphere, and southward in the southern hemisphere.

But the poleward distribution shifts that I've described so far occurred over thousands of years—do we see similar trends over much shorter time frames that are relevant for modern-day climate change?

The British seaside

An interesting place to begin looking for poleward distribution shifts in more recent times is in the oceans. Oceans play a crucial role in the global climate system by storing and transporting enormous amounts of heat. In fact, the ocean's heat capacity—its ability to absorb heat—is about a thousand times greater than that of the atmosphere, so the oceans have absorbed much more heat than the atmosphere in recent decades.[6] Therefore, we should expect to see significant biological impacts in the marine realm.

As is the case with atmospheric temperature, changes in ocean temperature can vary considerably between regions, but the general global trend over the twentieth century was one of warming. What's more, this warming has been accompanied by changes in ocean salinity (mostly due to inputs of freshwater from melting ice caps and changing rainfall patterns), rising sea levels (largely because water in the oceans has expanded as it warmed), and increased acidity (because oceans are soaking up more carbon dioxide, which creates carbonic acid). This all adds up to a significant change in conditions for the biodiversity that inhabits the world's oceans and shorelines.[7]

Observations of changes in temperature are much scarcer for the ocean than for the land, but the coast of Britain is one of the few places on Earth with a decent record, spanning multiple decades, of both water temperature and marine organisms. Observations of sea temperatures around Britain date back at least to the mid-1800s, but careful, regular sampling began in earnest in the 1880s and 1890s when seaside resorts vied with one another to claim warmer—or, perhaps more realistically, not so cold!—water to attract visitors.[8]

Records show that sea surface temperatures around the British coast vary considerably from year to year. But if we look behind these annual fluctuations, some general trends are apparent. From the early 1900s until the late 1950s, sea surface temperatures gradually crept up, increasing by almost 1°F over the half-century. However, the trend reversed in the early 1960s as temperatures began a steady decline, and by the early 1980s the sea was nearly as chilly as it had been around the turn of the century. Then, the trend reversed once again, and sea surface temperatures have been increasing since the early 1980s substantially faster than at any time in the previous 100 years (temperatures increased by nearly 2°F during the 1990s).[9] This flip-flop in regional sea temperatures—comprising a period of warming, followed by cooling, and then warming again—provides an ideal opportunity to explore how biodiversity responds to changing conditions over time frames of just a few decades.

Many organisms that inhabit the coastline around Britain are either warm-water species that are rarely found farther north, or are cold-water species that are rarely found farther south. Let's take barnacles as an example. Barnacles mostly come in two types around the UK: the *Chthamalus* species, which are characteristic of warmer waters and are distributed south as far as the tropics; and the species *Semibalanus balanoides*, which is mostly found in colder waters and penetrates far north into the Arctic Circle. The two types are only found side by side around the British Isles and northern France.

Surveys of these barnacle species along the southwestern coast of England date back to the 1950s, at which time *Chthamalus* were most common (numbers were especially high following the very warm years of 1958 and 1959). However, as sea temperatures cooled through the 1960s and 1970s, the cold-water species *S. balanoides* rapidly became more abundant. Then, when sea temperatures began to increase in the 1980s, *S. balanoides* began to retreat back to cooler waters farther north, while *Chthamalus* populations started to expand again and are now as common as they were in the 1950s.

This shifting dominance between species clearly reflects changes in sea temperature, but the precise mechanism by which this occurs is intriguing. It turns out that both species can tolerate a fairly wide range of temperatures—they are both able to reproduce and feed in harsher conditions than those in which they are usually found. However, space on rocks around the shoreline is limited, so the important factor is not so much whether each species can withstand the sea temperature, but whether each species can out-compete the other.

This is where temperature becomes important: studies have shown that *S. balanoides* reproduces more efficiently below about 59°F, while *Chthamalus* is more successful above this threshold.[10] So during cool periods, *S. balanoides* has a reproductive advantage and dominates the shoreline, but as the temperature rises, *Chthamalus* gains the upper hand and forces its adversary into retreat.

Barnacles thus illustrate how species can respond to changes in temperature over relatively short time frames. Similar trends have been observed over recent decades for many organisms found along the seashore, with species extending their ranges northward during periods of warming, and southward during cooling. Here's a sample of just some of the warm-water species whose ranges have expanded around Britain since ocean temperatures began to warm in the 1980s: Flat Top Shell, Toothed Top Shell, China Limpet, Black-Footed Limpet, Small Periwinkle, Acorn Barnacle, Montagu's Stellate Barnacle, Poli's Stellate Barnacle, and the Strawberry Anemone.[11] Meanwhile, cold-water species, including the Common Tortoiseshell Limpet and Winged Kelp, have retreated farther north. In the most extreme cases, latitudinal shifts of nearly 200 kilometers have occurred.

Moving now into deeper waters, similar trends have also been observed among commercially important fish species. A study led by John Reynolds and Allison Perry of the University of East Anglia in Norwich, UK, has investigated how recent changes in sea temperature have affected distributions of a variety of species that are important for Europe's fishing industry, such as cod, herring, and haddock, as well as some fishes of lesser market appeal, including bream and pilchard.[12] The Norwich team analyzed catches made by research vessels in the North Sea between 1977 and 2001.

They found that two-thirds of the 36 species studied shifted their distributions toward cooler waters as the sea warmed during these years. The majority of these shifts were poleward, as expected, with the distances moved ranging from about 50 to 400 kilometers. However, some species responded to the changing conditions by moving into deeper—and therefore cooler—water. Sinking down to cooler water is a response in the marine realm that is analogous to the upslope shifts on terrestrial mountains that we saw in Chapter 2.

The Norwich team's results are unlikely to be explained by factors other than temperature change. In particular, pressure from fishing does not explain the trends, because the spatial distribution of fishing efforts has not changed in recent decades, and, if anything, total fishing pressure may have declined slightly during this period. What's more, we can look back earlier than the Norwich team's study and find evidence that North Sea fish responded to the temperature swings prior to 1977. Longer-term data collected by the Marine Biological Association of the United Kingdom shows that cold-water species were gradually replaced by warm-water species during the first half of the twentieth century, and that this trend reversed in response to the cooling of the 1960s and 1970s. Interestingly, the 1960s and 1970s saw a marked upturn in the regional fishing industry, since fish that are well suited to the British market—including cod, herring, and haddock—are mostly cold-water species.[13] It seems the British palette is best adapted to cooler temperatures.

Continued increase in sea temperature over the twenty-first century is therefore likely to cause trouble for the fishing industry. Based on recent trends, the Norwich team estimates that two types of commercially important fishes—blue whiting and redfishes—may retract completely from the

North Sea by the 2050s. Given that fish populations are already under tremendous pressure from fishing, marine biodiversity in the region is likely to be significantly affected over the coming decades.

Bird watching in North America

Birds are especially good subjects for investigating distribution shifts. One key reason for this is that the ability to fly brings with it the potential to move easily around the landscape. Birds therefore have an obvious advantage over other land animals in their ability to track suitable conditions as the climate changes. Birds are also particularly useful for our investigations because excellent records of sightings over multiple decades are often available. In particular, notebooks from amateur bird enthusiasts can provide vital information for scientific research.

Researchers Alan Hitch and Paul Leberg from the University of Louisiana have analyzed data collected by thousands of volunteer birders in North America.[14] The information studied by Hitch and Leberg is collated in the North American Breeding Bird Survey, which is a huge effort to monitor the status of the continent's bird populations. Since the 1960s, more than 3,000 roadside routes in northern Mexico, the United States, and southern Canada have been repeatedly surveyed by volunteers who are trained to use rigorous methods for identifying and recording bird species. The result of half a century of effort is a phenomenal dataset tracking populations of more than 400 bird species.

However, a limitation of the Breeding Bird Survey for studying climate change impacts is that it encompasses the entire distribution of very few species; that is, the northern and southern edges of a species' range can rarely be assessed simultaneously. This is an issue because it makes it difficult to distinguish cases in which a species' distribution is shifting poleward (meaning that both northern and southern range edges shift north) from cases in which a species is expanding or contracting its range (the northern and southern edges shift in opposite directions). There are a number of explanations unrelated to climate change for why birds may be expanding or contracting their ranges—including changes in agricultural practices that favor birds, or destruction of nesting sites that lead to population declines—so it is crucial that we are able to tell these trends apart from poleward shifts.

To address this, Hitch and Leberg divided the surveyed bird species into those that had southern and northern distributions: any species that did not extend south of the latitude of (roughly) Atlanta were classified as having a northern distribution, and any species that did not extend farther north than the latitude of Toronto were classified as having a southern distribution. This classification allowed for the detection of shifts in the northern edge of the "southern" species, and in the southern edge of the "northern" species. If climate change is affecting bird distributions in North America, we would expect to see poleward shifts at both northern and southern range edges.

Hitch and Leberg also had to contend with a number of other factors that could have confused their analyses. Some species were excluded from the study because their distributions are known to have been directly influenced by human introductions (for example, game birds). Species dependent on lakes and rivers were also rejected because their distributions are unlikely to have been adequately surveyed by the roadside localities used in the Breeding Bird Survey. Finally, the entire Rocky Mountain region was excluded from the study because of the potentially confounding effect of upslope distribution shifts: species in mountainous regions are more likely to respond to climate change by moving upslope, rather than to new latitudes.

In all, 55 bird species satisfied Hitch and Leberg's stringent criteria for inclusion in their study. So what did these species show? Sure enough, the southern birds generally shifted their northern range edges northward, by an average of more than 2 kilometers per year. A few species bucked the trend by moving southward, but the clear overall trend was one of northward shifts at northern distribution edges. In some cases, the shift was very substantial. For example, the Inca Dove—which ranges from the southwestern United States down through Mexico and as far south as Costa Rica—extended its distribution northward by roughly 250 kilometers since the late 1960s; similarly, the Great-Tailed Grackle, a large blackbird distributed from Southern California down to Peru, shifted northward by more than 300 kilometers.

What about the southern range edges? In this case, Hitch and Leberg found no clear trend—some northern species shifted their southern range edge northward and some shifted southward, but there was no overall drift in either direction. That's not the northward shift that would be expected due to climate change, neither is it a southward shift indicating overall range expansion. (We'll return to ask why the southern range edges might not have shown northward movement in a moment.) But despite this, the overriding trend across all species—both northern and southern—remains that of a northward expansion: northern edges of distributions were moving northward, while southern edges showed no consistent trend.

A crucial part of this story is that the finding is very similar to a previous study in Great Britain. In fact, Hitch and Leberg based their analyses on a study undertaken by British biologists Chris Thomas and Jack Lennon in the late 1990s.[15] Thomas and Lennon used a similar survey of roughly 100 British bird distributions dating back to the 1960s and concluded that the northern edges of many species' distributions have moved northward (by an average of roughly 1 kilometer per year), while southern edges did not show any systematic shift. This consistency in findings across two continents and more than 150 species makes it very difficult to implicate any factor other than climate change as the cause of poleward distribution shifts in birds.

So why didn't southern range edges on either continent show a northward shift? A similarly perplexing result is common in studies of upslope distribution shifts, where expansions at higher elevation distribution edges are much more frequently observed than upslope contractions at lower elevation edges. Chris Thomas—whose extensive body of work we'll draw upon numerous times throughout this book—has a good, and remarkably simple, explanation for why this might be.[16] The key is to notice a quirk in how the survey data is generated. Surveys record that a species has moved into a new area as soon as individuals are spotted, meaning that the establishment of only a small population is sufficient to document a range expansion. In contrast, movement *out* of an area will not be recorded until the density of individuals has dropped to a level such that the species is not detected during repeated surveys. In conclusion, it is much easier to detect a range expansion than to confirm a range contraction. We should therefore not be surprised to find that poleward (and upslope) distribution shifts are more commonly observed at the poleward (or higher elevation) edge of the distribution, rather than at the equatorward (or lower elevation) edge.

The birds of North America and Britain therefore provide good evidence of highly mobile species rapidly shifting their distributions over large areas in order to track recent climate change. However, in the following section I will describe the predicament faced by a species that is much less mobile. This story shifts our focus to southern Africa. Take note that we're about to cross the equator, so all our thinking about the direction of poleward shifts in North America gets turned on its head: "poleward" now means to the south.

The Quiver tree of southern Africa

Quiver trees grow to around 4½ meters in height and are a distinctive feature of barren, rocky deserts in Namibia and South Africa. They are exceptionally well adapted to a hot climate, with branches that are covered in a thin layer of white powder that helps reflect away the sun's rays. The Quiver is also a succulent, meaning that it retains water in its leaves and stems, similar to cacti. The tree's crown is rounded and sits on top of the trunk rather like a golf ball set on a tee. This rounded crown formation is the result of repeated splitting of the branches, earning the species its scientific name, *Aloe dichotoma* (dichotoma meaning "split," or "forked"). The common name, by comparison, comes from the fact that Bushmen used to create quivers for their arrows from its branches.

Aside from this traditional use by humans, Quiver trees also provide important resources for a host of desert animals. Many insects are drawn by the copious nectar of the bright yellow flowers, and weaver birds often entangle their large haystacklike nests among the Quiver's branches. Quiver trees are therefore an important component of the desert ecosystem. Moreover, each tree maintains its role in the system for a long time, with life spans in excess of 200 years. Such longevity means that the species is slow to reproduce, with an estimated 15 years required for a newly established population to mature and produce seeds that can be dispersed to populate another area. This raises important questions as to the potential for Quivers to shift their distribution in response to climate change.

Botanist Wendy Foden took up the challenge of investigating the plight of the Quiver tree after hearing anecdotal reports that many populations seemed to be dying.[17] With support from a team of collaborators, Foden assessed populations at 53 sites spread throughout the entire range of the species. This was a formidable task considering that Quivers are distributed over an area of roughly 200,000 square kilometers, which is approximately equal to the combined size of England and Scotland. At each site, Foden and her travel companions carried out a fairly detailed health check on the local population of Quivers, measuring things like the fraction of leaves that had been shed and the amount of disease. Most importantly, at each site the percentage of dead trees was determined so that levels of mortality could be compared across the range.

Back behind a computer at the South African National

Biodiversity Institute in Cape Town, Foden analyzed her field data and found a clear trend of decreasing mortality moving from the northern to the southern ends of the species' distribution. Statistical tests proved that high rates of mortality (up to 70 percent of the population) were more common toward the northern edge of the range rather than toward the southern edge (where rates were as low as 2 percent). Further statistical wizardry enabled Foden and her collaborators to show that there was also a robust elevational trend in their data: populations at low elevations tended to experience more mortality than those at higher elevations. These findings, of course, concur with what we'd expect to see due to the impact of climate change, with populations experiencing more stress toward the warmer edge of the range than at the cooler edge.

Another line of inquiry that led to similar conclusions involved the use of old photographs, taken as long ago as 100 years. The researchers scoured archives for old photographs of landscapes dominated by Quiver trees, then traveled to where the photos were taken and took exact modern-day replicates. This industrious approach enabled Foden's team to estimate the rate of population growth or decline in different parts of the range. And, sure enough, once again populations were found to be in better shape toward the cooler, poleward edge of the distribution.

Foden and her colleagues investigated all possible explanations they could think of for the observed patterns in Quiver mortality—including exposure to livestock, pollution damage, and disease outbreaks—but the link with climate change seems unassailable. A close look at temperature records for this region reveals warming of at least 1°F over the last half century, and adding precipitation records reveals that the quantity of water available to plants has dropped throughout the Quiver's range (largely because increased temperature leads to more evaporation). Water availability is especially crucial for succulent desert plants, so the lack of moisture is sure to have put a strain on Quiver populations. Indeed, Foden's team noticed telltale signs during their travels that Quiver populations were suffering from a lack of water. Populations with high mortality were found to also have a high proportion of living trees that were shedding their leaves. This is an expected response of succulent plants to long-term water stress—the leaves simply dry out, wither, and drop. It is a process sometimes termed "auto-amputation," because after shedding, the branches form stumps that will never resprout.[18]

So we have strong evidence that climate change is exerting pressure to shift the distribution of Quiver trees poleward. However, unlike with the previous examples we've looked at in this chapter, the Quiver tree is not showing signs of shifting its range. The long life cycle of this species means that it is unable to colonize new areas fast enough to keep pace with the changing climate. Over time, the Quiver's distribution will be gradually squeezed as its northern edge contracts due to population declines, while the southern edge fails to expand. This big squeeze is bad news for Quiver trees, and it's also bad news for the insects and birds that rely on these immobile giants as oases of life in the hostile desert.

It is often assumed that desert species are less vulnerable to climate change because they are well adapted to high temperatures, but we're seeing that this assumption is unfounded. And what's more, there is much more biodiversity to lose in desert environments than you might have thought. The region over which Quiver trees are distributed—called the Succulent Karoo, after its most abundant type of plant—has been designated as a biodiversity hotspot[19] and contains more than 5,000 species, of which roughly 40 percent are endemic to the region. It also contains roughly a third of the world's succulent plants. If Quiver trees are representative of other succulents, then it seems there is the potential for climate change to cause high rates of extinction among desert species.

CHAPTER 4

Out of Sync

Waiting for spring

There is a long history of people recording the first signs of spring—the first daffodil to emerge, the first migrating swallow to arrive, the first singing thrush. Such records seem to be most commonly kept in cold regions, where signs of spring indicate that the long winter has finally passed. One of the longest records of the onset of spring is from the Marsham family estate in Norfolk, England, where Robert Marsham, a wealthy landowner, began recording the dates of springtime activities for more than 20 species of plants and animals in 1736. His record was maintained more or less continually by subsequent Marsham generations up until the late 1950s. Perhaps not surprisingly, the Marshams and other avid record-keepers have tended to show less enthusiasm for recording the onset of fall and the end of summer.[1]

My focus in this chapter is the impact of climate change on *phenology*—seasonal events in a species' annual life cycle. Plants and animals exhibit seasonal patterns in their activities because there is often only a limited period in the year when the environment is favorable for successful reproduction or growth. The activities that are most demanding for a species—laying eggs, feeding young, producing pollen—are best undertaken when conditions are optimal. Many annual events in a species' life cycle are therefore closely linked with the climate, which provides cues to signal when it is a good time to set about mating, spawning, feeding, flowering, emerging from hibernation, or migrating.

We could start our search for impacts of climate change on phenology with the Marsham record (which does indeed reveal some signs of recent climate change)[2], but I will instead begin with an example from Wisconsin, where a remarkable set of phenological records has been collected at the family farm of famed environmentalist Aldo Leopold. Leopold collected the first set of these records from 1936 until 1947. The records halted following his death, but a second set of observations, spanning the period 1976–98, was subsquently added by his eldest daughter, Nina Leopold Bradley. Between them, father and daughter recorded 74 different indicators of spring over a total span of 61 years, giving us a rare insight into how phenology has changed over time. They recorded indicators including arrival dates for migrating Canada Geese, American Robins, Wood Thrushes, and Eastern Meadowlarks, and dates of first bloom for Wild Geraniums, Field Pussytoes, Forest Phlox, and Marsh Milkweeds.

In continuation of the Leopold family's contribution, Aldo's youngest son, A. Carl Leopold, joined his sister as co-author of a scientific paper that presented this set of records in the late 1990s.[3] Overall, the data revealed an advance in springtime phenology since the 1930s: on average, spring events were occurring roughly one day earlier every decade. In total, 19 of the indicators that were measured revealed a clear trend toward earlier occurrence that could be shown to be statistically reliable, meaning it is very unlikely we are imagining a pattern that isn't really there. By contrast, none of the indicators revealed a reliable trend toward delaying of springtime phenology, providing strong evidence that spring is now arriving sooner at the Leopold's farm than it did in the 1930s and 1940s.

This advancement in spring phenology follows what we would expect due to twentieth century climate warming: as temperatures rise, conditions that trigger spring life-cycle events will tend to arrive earlier in the year. This is yet another fingerprint of climate change.

But by averaging across indicators—as with the "one day every decade" figure quoted above—we miss an important point: there is considerable variation in response between species. For example: the first Cardinal song advanced by about 3½ days per decade; Whip-poor-will, a species of nightjar whose name imitates its distinctive call, arrived on the farm nearly 2 days earlier per decade; the first bloom of Canadian Anemone was recorded almost a day and a half earlier per decade; and Common Milkweed bloomed a little more than 2 days earlier per decade. It is also important to realize that not all of the indicators measured in Wisconsin advanced their timing—in fact, about one third of the indicators have almost certainly not changed, with dates staying more or less constant during the period of study.

The reason that springtime has not come earlier in all cases is that many phenological events are not, in fact, regulated by climate. Instead, many life-cycle events are controlled by *photoperiod*—the number of hours of daylight per day. Because climate change does not have any affect on day length (planetary movements will continue to ensure that days get longer at the same time each year), we should not expect to see any adjustment in phenological events that are controlled by photoperiod. It is therefore likely that some of the non-changing indicators recorded by the Leopolds—including arrival dates for Eastern Bluebirds and Fox Sparrows, and flowering of Carolina Roses—are for species that do not rely on climatic cues for initiating their springtime activities.

We can take two important points from the Wisconsin study: first, there is a general trend for advancement in spring phenology that is consistent with global warming; and second, spring activities have advanced by differing amounts, or, in some cases, have not advanced at all.

These general conclusions from Wisconsin are supported by many other studies that have shown changes in phenology in different parts of the world. For instance: in Germany, apple and cherry trees are blossoming earlier; in Estonia, pike and bream are spawning earlier; in Britain, dozens of species of plants have advanced their flowering time; in Australia, migratory birds are arriving at their breeding grounds sooner; in Alberta, Canada, flowers are blooming earlier; and in Alaska, Pink Salmon are migrating earlier than in previous decades. And you might also like to know that the phenology of grapevines in Bordeaux, France, has advanced in the last two decades, resulting in merlot and cabernet sauvignon grapes that have higher sugar-to-acid ratios, which makes for better wine.[4]

The end of summer, from space

As I've already mentioned, most phenological records pertain to the onset of spring. There is also a geographical bias, with the majority of records that have been incorporated into Western scientific literature coming from Europe, North America, and, to a lesser degree, Australia. However, Xiaoqiu Chen and his colleagues at Peking University have used an ingenious way to study phenological trends in China—undeterred by the lack of records on the ground, they have used satellite data from space to look for trends in recent years. And a particularly neat thing about their approach is that they cannot only detect the onset of spring, but also the end of summer.

Satellites provide an amazingly efficient way to monitor the Earth's surface. By analyzing the spectrum of light coming from the planet—including, but not limited to, those wavelengths of light that we can see with the naked eye—it is possible to document the "greenness" of vegetation over huge areas. The onset of spring can be identified by looking for a burst of green in satellite images, while the end of summer is characterized by a change to brown as leaves turn autumnal. It is not usually feasible to detect changes for individual species, but the approach can give a good idea of when the growing season begins and ends across plant communities as a whole. Using images from NASA's archives, covering the period 1982–93, Chen and his colleagues were able to look for trends in the length of the growing season across a large portion of eastern China.[5]

So what is the message from space? As expected, the satellite images showed that the springtime start of the growing season gradually advanced in eastern China during the 1980s and early 1990s. The rate of change was a little under half a day per year, which is not too dissimilar to that seen in Europe and North America. However, much more noticeable in China was a delay of about one day per year to the end of the growing season, as indicated by the change from green to brown. The overall trend in China has therefore been a lengthening of the growing season, by a little less than 1½ days per year (or 15 days per decade).

This delay to the end of summer is more substantial than what has been observed in studies from Europe and North America. To understand why trends in China might be different from those seen in other parts of the world, we need to consider the region's unique climate. A detailed description of China's climate system is not necessary here, but suffice it to say the difference in heat reserves between the planet's largest continent and its biggest ocean makes for a distinctive regional climate, as is clearly demonstrated by the annual monsoon rains. It turns out that temperature records from China for the 1980s and 1990s show that the months of December through March have become warmer, while the months of April through June have become cooler. Chen and his colleagues have concluded that warming in late winter and early spring has triggered an earlier onset of the growing season (as we've seen elsewhere in the world), but more significant is that cooling during the late spring and early summer is delaying the end of the growing season, most likely because it is taking organisms longer to complete their life cycles in cooler conditions.

So we can take two key messages from the opening parts of this chapter: first, we have seen from the Leopold's farm in Wisconsin that different species are adjusting their life cycles by different amounts (or not at all); and second, the example from China demonstrates that responses in phenology are dependent upon regional patterns of climate change, meaning that responses will not be the same in different regions of the globe. Next I'll describe how these two issues can cause the phenology of species that are closely dependent on one another to fall out of sync.

Birds too late for caterpillars

Springtime in oak woodlands throughout much of Europe is characterized by the loud, melodious song of migratory birds such as the Pied Flycatcher, which is slightly smaller than a sparrow, with a black back, white underside, and black Batman-style head mask. The Pied Flycatcher escapes European winters by flying thousands of miles south to the warm forests of West Africa. Then, every spring, when the days get longer, the birds head back to the temperate forests of Europe to breed. Males normally undertake the long journey to their breeding grounds a little earlier than females because, on arrival, females select their mate for the season based on the quality of the male's territory. So it pays for the males to arrive in time for some extra housework, and their singing to attract arriving females onto their property is a typical sign of the onset of spring.

In addition to taking insects in flight (as its name suggests), the Pied Flycatcher feeds on caterpillars among oak foliage. Caterpillars emerge in the springtime and are an especially valuable resource for feeding newly hatched flycatcher chicks. In fact, the fairly short window between when caterpillars hatch and when they vanish to transform into butterflies and moths is crucial for flycatchers: they need to synchronize their breeding so that the time when their offspring are most in need of food coincides with the time when there is a plentiful supply of fat, nutritious caterpillars. Many insectivorous forest birds synchronize their life cycles with the availability of food, so by looking at the impact of climate change on the Pied Flycatcher, we can get an idea of the kinds of impacts that might be happening in other species.

Pied Flycatchers usually build nests in natural tree holes, but they will readily set up house in artificial nest boxes, making them ideal subjects for study by biologists who want to keep a watchful eye on what goes on inside the nest. Two such researchers—Christiaan Both and Marcel Visser—have studied populations of these birds in the Netherlands, where the majority of breeding pairs settle in nest boxes. In particular, they have studied the timing of phenological events at one of the largest and oldest nature reserves in the Netherlands, Hoge Veluwe, which is situated about an hour's train ride east of Amsterdam. Over the period 1980 to 2000, Both and Visser have documented an advance of about 5 days per decade in the date when Pied Flycatchers lay their eggs at Hoge Veluwe. Local temperatures in early spring have also crept up over these two decades, giving us yet another example of an advance in spring phenology that is apparently due to climate change.[6]

But what about the caterpillars? First, it is interesting to note the way that scientists measure the abundance of such tiny critters, which aren't as easy to count as birds in nest boxes —they weigh their droppings. This is not one of the most glamorous jobs in science, but by spreading a cloth at the foot of a tree, sorting through the debris that falls, and then weighing the droppings that are collected, it is possible to get a fairly accurate idea of the number of caterpillars.

Marcel Visser is currently leading a research project at Hoge Veluwe that has collated a long dataset of these measurements, taken every 3 or 4 days over a 25-year period beginning in the mid-1980s. This remarkable record reveals that the timing of peak abundance of caterpillars has also advanced as the climate has warmed. However, the rate of advancement, which clocks in at around 7.5 days per decade, is considerably faster than the rate at which Pied Flycatchers have advanced their egg laying.[7]

The worry, therefore, is that flycatcher chicks might increasingly hatch too late to coincide with the peak in food availability. By extending their study to incorporate nine additional oak woodlands spread across the Netherlands, Both and Visser have uncovered evidence indicating that this is indeed occurring. They found that flycatcher populations have declined by as much as 90 percent in woodlands where caterpillar abundance peaks very early. However, flycatcher populations have declined only slightly at sites where the advance in peak food availability has not been so substantial. It seems that the birds are unable to cope with large shifts in caterpillar phenology, though they can still manage when the shift in food availability isn't too great.[8]

In order to understand why there are limits to the fly-catcher's ability to advance its breeding, we need to look more closely at the bird's migration habits. In another case in which "citizen scientists" have provided crucial data for analysis by professional scientists, Both and Visser were fortunate to have access to the records of a local birding group. Since the early 1980s, this group has recorded the first date each year when the flycatcher's distinctive warble has been heard at Hoge Veluwe. Analysis of the data revealed that the birds have not been arriving earlier as the climate has warmed. So, although there has been a clear advancement in the phenology of their main prey, and in their own egg laying, the birds have not advanced their spring arrival in the Netherlands.

The advancement in egg laying, then, has been achieved by shortening the interval between arrival and breeding. In the past, the birds have arrived with enough time to respond to natural annual fluctuations in the start of spring—in a warm year, they must get down to business quickly, while in cooler years, they are best waiting a little while. But under present conditions, the interval between arrival and optimum breeding time has shrunk to only a few days, and further advancement of breeding is not possible without earlier arrival.[9] The window of opportunity has become too narrow and, in those areas where caterpillar phenology has advanced the most, flycatcher populations are unable to advance egg laying sufficiently to keep up, leading to the substantial population declines that have been observed.

So why haven't Pied Flycatchers advanced the timing of their spring migration? Well, we don't know for sure, but the examples we've looked at so far point toward two likely explanations. First, it is possible that the timing of migration in this species is linked to photoperiod, not climate, so their cue to leave Africa and head to Europe is unaffected by climate change. And, second, differences in regional patterns of climate change may mean that conditions at the time of departure from Africa no longer provide a reliable indication of what conditions will be like on arrival in Europe. What these birds lack when they set out are accurate weather forecasts for conditions at the end of their migration route.[10]

Although it is unknown which of these two explanations apply to migration phenology for Pied Flycatchers, it is very likely that both come into play for different species and in different regions of the world. We can conclude that numerous long-distance migrants are likely to suffer population declines under climate change due to mistimed arrival at their breeding grounds.

So we have evidence that a migratory bird is falling out of sync with its main food source, but this is just one link in a longer chain of interactions. If we look further down the food chain, we find additional synchronized events, each of which has the potential to become unhinged and to affect life cycles higher up the chain. So let's now take the story one link lower and ask: are caterpillars themselves falling out of sync with resources on which *they* depend?

Caterpillars too early for oaks

One of the principal types of caterpillar on which Pied Fly-catchers feed is the larvae of the Winter Moth, a common species throughout oak forests in Europe. The caterpillars grow to about 2½ centimeters in length and are green with white racing-style stripes along their sides. Adult Winter Moths emerge from the pupal stage (during which they undergo metamorphosis from larvae into adults) in early winter and hastily mate at the base of tree trunks. Females are wingless and so have a grueling climb up into the tree canopy where they lay their eggs. The eggs then enter a state of dormancy through the rest of winter, before the onset of spring triggers the moth larvae to hatch and begin feasting on canopy leaves. After they have gorged for a few weeks and massively increased their body weight, the larvae complete the Winter

Moth life cycle by dropping to the forest floor and burying themselves, to emerge as adults once the next winter begins.

A crucial part of the Winter Moth life cycle is the timing of egg hatching. Much like the Pied Flycatchers that feed on them, it is essential that caterpillars hatch at the time when their food source is at its peak, which in their case means that they must emerge in synchrony with tree bud burst—the stage when leaves emerge from buds and start to unfold and separate. Winter Moth caterpillars that hatch before bud burst will starve, while those that emerge too long after bud burst will be faced with older leaves that are less digestible because they contain high tannin concentrations (much like red wines with high tannin content can be more difficult for us to enjoy). Thus, hatching either too early or too late results in decreased nutritional intake and reduced chances of surviving to adulthood.

Marcel Visser and his team have uncovered evidence suggesting that this link in the food chain is also under threat from climate change: advancement of Winter Moth phenology is too rapid, meaning that eggs are increasingly hatching before the date of bud burst.[11]

The key to understanding this example is to focus on the cues that each species uses to trigger hatching and bud burst. By studying records from the Netherlands along with additional data from Scotland and the United States, Visser and his colleagues were able to pin down exactly how each species responds to the environment. For oaks, bud burst is triggered each year when the tree has experienced a certain number of days with mild temperatures, above around 39°F. This is a fairly robust signal that winter is over and springtime has arrived. Winter Moths respond somewhat similarly, but with an important twist. It turns out that egg hatching is cleverly adapted to take into account the number of mild days (as with bud burst) and also the number of very cold days (the number of days during winter that drop below freezing). Basically, in years with cold winters, spring temperatures need to increase for fewer days before hatching occurs, while in years with relatively warm winters, spring temperatures need to increase for more days before hatching occurs.

Close examination of temperature records for the Netherlands reveals that spring temperatures have increased over recent decades, yet the number of very cold days during the winter has not substantially changed. This combination—cold winters and warm springs—is the ideal recipe for accelerating Winter Moth phenology, and most likely accounts for the extra rapid advance in caterpillar phenology that we've seen. Overall, it seems that this advance is causing mistiming across at least two links in the food chain: birds to caterpillars, and caterpillars to oaks.

Of course, if caterpillar populations crash because they are hatching before bud burst, then this will be more bad news for Pied Flycatchers, illustrating how mistiming at one point in the food chain could set off a chain reaction that leads to impacts further up. In the next section we'll explore how mistiming of phenology can resonate across an entire ecosystem, comprising a larger number of species than we have looked at so far.

A marine food chain

Let's now turn our attention back to an environment that we looked at in Chapter 3—the cool waters of the North Sea. First, it is necessary to understand some basics about marine food chains.

Right at the bottom of any food chain are the *primary producers*—those species that make organic material from inorganic compounds, usually through the process of photosynthesis. Primary production on land is carried out mostly by the higher plants (trees, shrubs, and flowering plants), but in marine environments the job is done chiefly by minute algae that drift with ocean currents. These microscopic life forms are termed *phytoplankton*, and two of the most common types are diatoms and dinoflagellates. Both diatoms and dinoflagellates are single-celled organisms, which are far too small to be seen with the naked eye. But when conditions in the ocean are favorable—meaning there is sufficient light, nutrients, and warmth—then the phytoplankton bloom and can be present in such high numbers that they discolor the water. Such blooms are common in spring and autumn, with smaller blooms throughout the summer. It is large blooms of dinoflagellates that are responsible for creating "red tides," which are well known because toxins produced by them can kill marine life and cause nasty illnesses in people who eat infected fish.

The next link up the food chain from the phytoplankton is the *zooplankton* ("zoo" referring to animals, as opposed to "phyto," which refers to plants). These tiny multicelled animals come in many different forms, but all are dependent on the primary producers for survival. One type of zooplankton is the copepods, which are crustaceans that typically grow to one or two millimeters in length. Most copepods feed directly on phytoplankton, making them *secondary producers*, and their phenology is therefore timed to take advantage of the spring, summer, and autumn blooms in their food. The result is a kind of succession within the marine ecosystem—first the diatoms and dinoflagellates bloom, then the copepods increase in abundance.

This succession continues as we move higher up the food chain. Next up are various, slightly larger species of zooplankton that tend to feed on the smaller zooplankton. These larger zooplankton include the shrimplike crustaceans called krill, which most commonly grow to one or two centimeters in length. The most important component of the diet of these animals is the copepods, making the larger zooplankton *tertiary producers* (phytoplankton being level 1; copepods being level 2; and krill and other large zooplankton being level 3). Krill are best known as the food source for many much larger animals, including (remarkably) the largest living species of all—the blue whale.

One more important link in the chain is fish larvae, which are components of the *meroplankton*—juvenile organisms that spend only part of their life cycle as plankton. Most fish have a larval stage (like the Winter Moths described in the last section) and during this phase it is essential that the larvae have plenty of copepods to feed on. This makes them tertiary producers and, ideally, the fish's larval stage will be timed to take advantage of the increase in copepods, which in turn follows the bloom in phytoplankton.

To investigate whether climate change is having an impact on phenology within this neatly synchronized system, marine biologists based in the British harbor town of Plymouth have analyzed reams of data collected by devices that record plankton communities in the water column.[12] These devices have been voluntarily towed behind roughly 100 merchant ships on their normal routes of passage since the 1930s, creating an extraordinary record of changes to plankton communities in the North Sea. To ensure their analyses were reliable, the researchers only used records since 1958 (because methods for recording have remained almost unchanged since then).

And to make the analysis of roughly 100 different types of phyto- and zooplankton more manageable, they assigned all species to five groups that reflect their position within the food chain: diatoms (primary producers), dinoflagellates (primary producers), copepods (secondary producers), larger zooplankton (tertiary producers, including krill), and meroplankton (secondary and tertiary producers, including fish larvae and other juveniles).

That is enough terminology—what did the reams of data show?

It turns out that, despite considerable fluctuations in the timing of peak abundance between years and between species, some significant underlying trends emerged. The timing of blooms in the late spring and summer has been getting earlier for all five functional groups. However, the rate of change was very different between groups. Over 50 years, the primary producers (diatoms and dinoflagellates) advanced by 23 days, the copepods and larger zooplankton advanced by only 10 days, and the meroplankton advanced by 27 days. Together, this adds up to substantial mistiming between successive links in the food chain—a disruption to the synchrony between primary, secondary, and tertiary production.

So we once again find evidence of shifts in phenology that vary substantially between species, and between different levels in the food chain. Not surprisingly, the annual timing for summer diatoms, dinoflagellates, copepods, larger zooplankton, and meroplankton has been shown to be strongly associated with sea surface temperature, so advances in phenology for these organisms are likely to be related to climate change.

The upshot of this disruption to the lower levels of the food chain is that we can expect to see important impacts higher up, on some of the largest animals in the sea, including the whales that feed on krill and the fish whose adult populations are dependent on the success of their larvae in the zooplankton. This chain of effects could well be having a significant impact on fish stocks in the North Sea. Cod and other commercially important fish species are reliant on the successful feeding of their larvae, and mistiming in the phenology of phytoplankton, zooplankton, and fish larvae is likely to reduce larvae survival. This implicates rising sea surface temperatures as a potential contributor to declining fish stocks in this part of the world.[13]

Climate change is certainly not the sole—or even the main—cause of dwindling fish populations, but it is reasonable to conclude that changes in sea temperature have exacerbated the effects of overfishing and other impacts. This potential for climate change to exacerbate the impacts of other threats is a recurrent theme throughout the book. For the North Sea, take the phenological mistiming we've heard about in this chapter, add the poleward distribution shifts of intertidal organisms and fish we saw in Chapter 3, and then add to the mix overfishing and water pollution, and you have a potent combination of factors threatening the entire system.

CHAPTER 5

A Global Fingerprint

Rainforests of the sea

The Chagos Archipelago is pretty much as far from anywhere as you can get on this planet. The sixty or so small islands that make up the archipelago sit in the middle of the Indian Ocean, about 1,600 kilometers east of the Seychelles, and 450 kilometers south of the Maldives. The archipelago is the remnants of a cluster of several mid-oceanic volcanoes that have subsided over millions of years, creating at least a dozen rings of reefs, each encircling a shallow, turquoise lagoon. Most of the islands are uninhabited—the exception being Diego Garcia, at the southeastern edge of the archipelago, which houses a joint U.S. and UK military facility—making most of Chagos the picture-postcard image of a tropical island paradise.

The islands of the Chagos Archipelago have a combined area of just 63 square kilometers, roughly the size of Manhattan.

But the total area including the lagoons and reefs is very much larger, at more than 15,000 square kilometers, comparable in size to the state of Connecticut. This expanse of shallow water in the middle of the Indian Ocean comprises about 2½ percent of the world's coral reefs and is home to a remarkable abundance and diversity of marine life. Coral reefs are one of the most biologically diverse ecosystems on Earth, often earning them the title "rainforests of the sea." It's been estimated that nearly one quarter of all marine fish species live on coral reefs.[1]

It is the corals themselves that form the underlying basis on which the rest of reef life thrives. Indeed, the Chagos Archipelago is, in effect, made of coral—it is the physical structure constructed by coral that makes Chagos an oasis of life in the middle of the deep ocean. Elsewhere, corals are responsible for building and maintaining massive oceanic landmarks such as Australia's Great Barrier Reef and the Bahamas' Andros Barrier Reef. So what exactly are these remarkable reef-building corals?

Although they tend to resemble plants, corals are, in fact, animals. On the evolutionary tree of life, they are closely related to anemone and jellyfish. Moreover, while corals are commonly perceived to be a single organism, the formations we see are actually created by colonies of thousands of tiny coral "polyps." Each polyp is itself an individual organism with a body rather like a miniature anemone, the bottom end holding fast to the rest of the colony and the top end sporting a mouth surrounded by tentacles. The polyps each secrete calcium carbonate to form a hard, protective skeleton, and over many thousands of generations these skeletons layer on top of one another to form the structure that we commonly perceive as a coral. Astonishingly, colonies of coral polyps develop in such a way that the overall structure has a distinct form that is characteristic of the species—take, for example, *staghorn* and *brain* corals, whose names describe their distinctive forms. It is these calcium carbonate structures that, over thousands of years, are broken down by waves into fragments that settle and amalgamate to form the structure of the reef.

Another remarkable thing about corals—something that is crucial for our point of interest—is that they live in a close, mutually beneficial relationship with a type of microscopic algae that sets up home within the coral's tissues. Although coral polyps are able to catch plankton by using their tentacles, corals receive the vast majority of their nutrients from these algal squatters. The algae photosynthesize and in doing so produce numerous compounds, including sugars and complex carbohydrates, that are absorbed by the coral as a source of food. And the benefits work both ways: corals in return provide the algae with essential plant nutrients, especially ammonia and phosphate, from their waste materials. So corals and algae have formed an extremely successful partnership—a relationship that is formally called *symbiotic*—with each largely dependent on the other for survival. So close is their association that algae are even responsible for giving corals their characteristic color.[2]

Like all the organisms we've looked at so far, corals are greatly influenced by the nature of their physical environment. Ocean temperature is one key factor in determining where reef-building corals can grow (other factors include salinity and availability of light), with reefs tending to thrive in tropical waters where temperatures are between 64 and 86°F. At cooler temperatures, corals tend to be outcompeted by forests of kelp and other temperate organisms, and at higher temperatures—well, now we're moving into largely uncharted waters.

We can begin to understand the effects of temperature

increases on corals by taking a look at the recent history of Chagos. Charles Sheppard, a marine biologist at Warwick University in the UK, led scientific expeditions to Chagos in the 1970s and 1990s. Undertaking one of the more enviable jobs in science, Sheppard and his colleagues surveyed the reefs by scuba diving and snorkeling around large sections. In doing so, the team documented an incredible diversity of life on the archipelago's reefs, including more than 200 species of coral and more than 700 species of fish. Crucially, by revisiting the archipelago over the course of two decades, the scientists were able to observe changes over time.[3]

Sheppard's expeditions found that in the 1970s, coral cover was extremely high on each of the reefs surveyed, with at least half to three quarters of each reef covered with living corals. By 1996, some deterioration in the quantity of coral had been noted, but overall the reefs remained fairly healthy.

Then something happened. By the time Sheppard and his team visited in 1999, there had been a massive reduction in the amount of living coral. Vast areas of pale, lifeless coral dominated the underwater landscape, and very few fish and other reef species remained. Little more than one tenth of the reefs still supported living corals, and Sheppard reported being able to snorkel for long periods without seeing a single live specimen in areas that were thriving just three years before. Many of the dead corals were at least 200 years old, so this was clearly not an event that happens frequently.

The extreme remoteness and pristine nature of Chagos makes it very unlikely that local factors, such as chemical pollutants or tourism, were responsible for the mass coral mortality observed in Sheppard's 1999 expedition. In fact, we can now be confident that the collapse of the coral population was caused by an increase in ocean temperature, which led to a breakdown of the all-important coral-algae relationship. It turns out that as ocean temperatures rise, algae tend to flee coral tissues resulting in a loss of color that is commonly termed "coral bleaching." While other threats, such as changes in salinity, cause death to the coral polyps themselves, bleaching is a telltale sign that raised ocean temperatures are to blame.

The time immediately after bleaching is critical for coral survival—if temperatures cool, the algae will recolonize the coral tissues, but if the algae deserts for too long, the coral will suffer from malnutrition and eventually die. Once the coral dies, the physical structure of the reef will support a diversity of marine life for a while, but without the glue that holds it together, the reef will soon begin to erode and may eventually be completely destroyed. Loss of the physical structure of the reef ultimately means loss of the many organisms that are part of reef life, and this in turn spells trouble for coastal societies in parts of the world that rely on the reefs for harvesting food and for tourism dollars.[4]

Unfortunately, there aren't any local records of changes in ocean temperature around Chagos, but weather data recorded at the military base on Diego Garcia confirms that this remote place has not escaped the global trend for warming temperatures, with a 1.8°F rise in air temperature recorded over the last quarter of the twentieth century. So the mass coral mortality at Chagos coincided with a warming environment, lending support to the argument that climate change was responsible.

However, to fully understand the role of climate change in causing coral bleaching, we need to broaden our perspective and take a look at what is happening worldwide.

Global bleaching

Because many societies live in close proximity to reefs and are dependent on them for food and income, any deterioration in reef health tends to be noticed promptly by the local population. In the last decade or so, global networks (such as the Global Coral Reef Monitoring Network)[5] have been established to collate observations of reef deterioration, meaning that the health of coral reefs around the world is fairly well monitored. These networks are increasingly recording massive bleaching events that affect corals over vast areas. For instance, in 2002 the Great Barrier Reef experienced mass bleaching across an area stretching at least 2,000 kilometers, and in 2005 reefs throughout a large portion of the Caribbean were hit hard.[6]

The years when Chagos suffered bleaching—1997 and 1998—were exceptionally bad years for corals in almost every tropical region of the world. The first signs of impending problems emerged in mid-1997, when warm waters pooled up in the far eastern Pacific, against the coast of Central America. In May, the first bleaching was observed at Malpelo Island, which lies 500 kilometers off Colombia's Pacific coast, and by mid-September, substantial bleaching had occurred northward through Panama, along the coast of Mexico and as far north as the Gulf of California.

By December 1997, warmer than usual waters had expanded out to the Galápagos Islands, where nearly all corals bleached to some extent over the following three months. Then, in the first half of 1998, bleaching occurred throughout the Indian Ocean (including the Maldives, Seychelles, and Sri Lanka), along the coast of Eastern Africa (notably, Kenya and Tanzania), in East Asia (Indonesia, Thailand, and Vietnam), and in Australia (including the Great Barrier

Reef). And it didn't stop there: the second half of 1998 saw bleaching spread into the Persian Gulf and the Red Sea (Bahrain, Qatar, and Saudi Arabia), the Far West Pacific (Micronesia), and the Caribbean (Bermuda, Barbados, the Caymans, and Jamaica).[7] The severity of bleaching varied substantially around the world, but in total, roughly 10 percent of the world's reefs were damaged to the extent that they had not shown significant signs of recovery by 2004.[8]

So, the coral mortality that was observed by Charles Sheppard at Chagos in 1999 most likely occurred in the first half of 1998, at the same time other islands in the Indian Ocean were suffering. Put in a broader perspective, the bleaching at Chagos was part of a massive wave of mortality that spread, literally, around the world.

Looking at all the observations together, we find a very strong association between bleaching and high water temperatures: bleaching coincided with a period of unusually warm waters that began in mid-1997 in the far eastern Pacific and spread around the globe until late 1998, shortly after the first typhoon of the season cooled the waters off Japan and Taiwan. This tight association between temperature and bleaching supports the notion that global climate change is the principal cause of mass coral mortality, especially when we consider that oceans today are warmer than they have been for at least the past 420,000 years.[9]

But hold on a moment! A skeptic is likely to object by calling on El Niño to explain the unusually warm temperatures of 1998. To address this objection and gain a deeper understanding of the issue, we need to first take a brief detour to look at El Niño and its counterpart, La Niña.

El Niño and La Niña refer to stages in a natural cycle of shifts in global oceanic and atmospheric conditions. The name

El Niño—"the boy"—was coined as a reference to the Christ child, because the phenomenon is usually noticed around Christmastime along the Pacific coast of South America, and the name *La Niña*—"the girl"—is commonly used to refer to the opposing part of the cycle. El Niño and La Niña stages occur at irregular intervals, tending to recur roughly every 3 to 8 years, when shifts in the intensity of the trade winds trigger a sequence of events that affect rainfall patterns and ocean temperatures. The kinds of weather that El Niño and La Niña bring vary according to where on the planet you live: for example, peoples of Indonesia, Australia, and southeastern Africa are likely to equate El Niño with severe droughts; whereas Ecuadorians, Peruvians, and Californians tend to associate it with torrential downpours. By contrast, La Niña tends to cause conditions that are opposite to those of El Niño; for instance, wet weather in Australia and southeastern Africa. In general, El Niño and La Niña conditions last between one and two years, and although some of the details of what triggers the cycle remain puzzling, we can be sure that this is a natural phenomenon that occurred long before humans started pumping vast quantities of greenhouse gases into the atmosphere.[10]

So might the El Niño–La Niña cycle be the true cause of the mass bleaching events of 1997–98? Here we have a classic example of the difficulties of distinguishing natural and human-driven affects, and the answer to our skeptic's question cannot be boiled down to a simple yes or no. The 1997–98 bleaching did coincide with one of the most extreme El Niño–La Niña events in recorded history. El Niño conditions were first noticed in early 1997, and by the middle of the following year had raised ocean surface temperatures in the eastern Pacific and Indian Ocean by as much as 7.2°F above normal.

Then, in a rather sudden swing, La Niña conditions took over in mid-1998, raising ocean temperatures in the South China Sea and west Pacific.[11] So El Niño and La Niña do seem to explain the 1997–98 bleaching event. Other mass bleaching events—occurring in 1987–88, 1993, 2002, and 2005—also correspond with strong El Niño stages.

However, the correspondence in recent years between bleaching and El Niño–La Niña misses a key point: bleaching events seem to be getting more frequent. Although there have been at least five incidences of mass bleaching since the 1980s, reports of bleaching prior to those events are extremely rare. There are very few earlier reports of bleaching from tourist resorts or from intensively studied scientific research stations. And underwater filmmakers who filmed extensively on the Great Barrier Reef during the 1960s and 1970s never saw coral mortality on the scale that has been seen in the last three decades. Thinking a little further back in time, it is interesting to note that some indigenous societies that are dependent on the reef for fishing (for example, in French Polynesia) do not have traditional terminology to describe mass bleaching. It seems unlikely that large-scale bleaching, resulting in hardship from a loss of fish stocks and other resources, would have gone unnoticed, and so unnamed, for many generations within such societies. A skeptic might argue that the apparent increase in frequency of mass bleaching events can be explained by the increase in the number of scientific observers, and the implementation of global monitoring systems, in recent years. Yet, overall, there is little reason to believe that bleaching events were as regular prior to the 1980s.[12]

The apparent increased frequency of mass bleaching events supports the theory that global climate change is at least partly responsible. There is no reason to expect increased frequency due to El Niño–La Niña cycles acting alone—these natural events have occurred for many years before widespread bleaching became so frequent.

Overall, it is most likely that the bleaching observed at Chagos and elsewhere has been caused by a combination of two principal factors: first, strong El Niño–La Niña events, and, second, a rising baseline of ocean temperatures that is linked with climate change. So bleaching occurs when El Niño–La Niña cycles bring peaks in ocean temperatures, but these peaks are overlaid on a general trend toward rising average temperatures, meaning that each time a peak occurs, it has a tendency to push temperatures a little higher than before. So once again, we find that climate change is an underlying factor that works in combination with an additional factor: climate change loads the dice against coral survival in years with strong El Niño and La Niña events.

Before we move away from our discussion of corals, I want to mention an additional threat that results from increased emissions of carbon dioxide, much of which are absorbed by the oceans (roughly one quarter of all the CO_2 emitted by human activities is absorbed by the oceans[13]). Increased availability of carbon dioxide is advantageous to species that photosynthesize—an effect known as "CO_2 fertilization." Indeed, many studies have demonstrated increased plant growth under elevated CO_2,[14] and anyone who has used a CO_2 injection system for their home aquarium will vouch for the explosion in growth of plants. However, a more ominous issue is that when carbon dioxide reacts with water, it produces carbonic acid, meaning that the world's oceans are gradually turning acidic. This, as you might guess, is not especially good news for the scores of marine organisms that build shells and skeletons from calcium carbonate, which is prone to dissolution in acidic water. Organisms such as corals and shellfish are consequently struggling to build and maintain their skeletons as oceans acidify.[15]

We therefore find that corals and the rich diversity of life that they support are at risk from a number of different threats, including water temperature rises and ocean acidification. A number of studies have predicted severe impacts on coral reefs over the coming century, with results consistently pointing toward eroding reef structures and fewer species. One fairly comprehensive study concluded that as many as one third of all reef-building corals are at risk of extinction due to the combination of climate change and other threats.[16]

Meta-analyses

Our search for biological fingerprints of climate change has taken us to many different parts of the globe—including Madagascar, Costa Rica, the North Sea, Southern Africa, North America, China, and the Netherlands—and has included a range of different organisms, including frogs, birds, caterpillars, barnacles, trees, and plankton. Each of the studies we've looked at so far has provided strong evidence that climate change is impacting species, but it is extremely difficult for any study from a single region, or for a limited set of organisms, to provide unequivocal evidence that climate warming is a key driver of change. Lingering questions tend to remain: Could the observed changes be caused by nonclimatic factors, such as habitat destruction? Could the changes be part of a natural cycle? In wrapping up our discussion on fingerprints of climate change, let us ask whether all the individual pieces of evidence collectively prove that climate change is already affecting living systems.

An important problem for biologists assessing climate change impacts is that records showing biological responses tend to be fairly short. There are exceptions—such as the 60 years of records from the Leopold family farm in Wisconsin —but the majority of available data spans little more than a couple of decades. Biologists are envious of climatologists in this regard, since records of temperature and rainfall tend to stretch back at least 100 years in many parts of the world. The relative brevity of biological records reduces our confidence in deciphering unnatural changes, since we are less able to place these changes within the context of longer-term fluctuations. However, biologists are a resourceful bunch and have managed to overcome the limitations of short timespans by exploring trends across many individual studies, using a technique termed *meta-analysis*.

The thinking behind meta-analysis is basically "strength in numbers": there is much more confidence in a conclusion if lots of pieces of evidence point in the same direction. So, what we first need to do is identify which kinds of impacts of climate change should point to the same conclusion all around the world.

It is no coincidence that the three main fingerprints that we've looked at so far—upslope distribution shifts, poleward distribution shifts, and advanced spring phenology—are those that are of most use to us now. I must emphasize that these three trends are by no means the only biological impacts that we are seeing from climate change. Other impacts that I haven't yet mentioned include changes in reproductive success (for instance, reduced breeding in polar bears)[17] and variation in body size (for instance, increased body size of some rodents in Japan has been linked with rising temperatures, most likely through increased food availability).[18] However, an important point is that some impacts of climate change can be expected to result in opposing trends in different parts of the world and for different organisms, whereas other impacts (in particular, the three fingerprints we have looked at so far) should show consistent responses all around the planet.

Take reproductive success as a brief example. Studies have shown that polar bears are suffering from reduced reproductive success under climate change, most likely because their body condition is deteriorating due to melting Arctic sea ice, which affects their ability to catch seals (more on this in Chapter 8). But in stark contrast, recent warming has been shown to *increase* reproductive success in some British butterflies because of increased availability of habitat suitable for egg laying.[19] So climate change may cause either increases or decreases in reproductive success, depending on the species. However, trends for the three fingerprints I have focused on so far should be consistent all around the world and for the vast majority of species.

Three large meta-analyses have been undertaken in the search for a global fingerprint of climate change, one by Camille Parmesan from the University of Texas and Gary Yohe from Wesleyan University, a second led by Stanford University researcher Terry Root, and a third headed by NASA scientist Cynthia Rosenzweig.[20] Between them, these three studies took on the Herculean task of assessing and summarizing observations from hundreds of research papers, including records from all continents and for thousands of species, ranging from mollusks to mammals and from grasses to trees. The exact methods used by the studies differed a little, but in each case, the general approach was to include any research paper that reported changes in species' distributions and/or phenology over a decent span of time (at least 10 years but, ideally, longer). Observations from the studies were then classified as being either consistent or inconsistent with a warming trend, enabling grand, overarching conclusions to be drawn.

Parmesan and Yohe identified 484 species whose phenologies had changed, of which 87 percent showed the expected trend toward spring advancement, and 434 species whose distributions had changed, of which 80 percent had shifted in the direction expected due to climate change (upslope or toward the poles).[21] Similarly, summarizing across shifts in distribution and phenology, Root's team concluded that more than 80 percent of species assessed were shifting in the expected direction, while Rosenzweig and her colleagues—whose analysis is most up-to-date—found the proportion to be up around 90 percent. So on the whole, shifts are overwhelmingly in the direction expected due to climate change.

And it turns out that the data tells a similar story when you look at it in multiple different ways. For instance, Rosenzweig and her colleagues compared patterns of temperature change across the globe against locations where studies in their analysis had shown biological changes. They found that those places where species are predominantly changing in the direction expected due to climate change are, in fact, the places that have experienced most significant warming. Because climate change is not uniform over the planet, with some areas warming much more than others, the observation that most biological change is occurring in places experiencing the most warming is further strong evidence that climate change is behind many of the changes we're seeing.

All in all, whichever way you look at the data, scientific

reports strongly indicate that a significant impact of climate change is already discernible among plants and animals. By the IPCC's reckoning, there is *very high confidence* that terrestrial ecosystems are already being impacted by climate change, and *high confidence* that marine and freshwater systems are being impacted.[22] The impacts are less sudden and immediately obvious than effects caused by other factors, such as habitat destruction, but we can be pretty sure that climate change is a threat that is quietly creeping up on us—a "stealth" threat, as I termed it in Chapter 1.

The evidence seems fairly unanimous, but it is important to think through any factors that could possibly bias our conclusions. One such bias could result if studies that support a role for climate change are more likely to make it into the published literature. This might occur if researchers are more likely to write up results for species that shifted in the direction expected due to climate change, or if journal editors and reviewers (who are responsible for selecting what results get published) preferentially select those results that are in agreement with climate change. These are not things that are easy to quantify, but it is reasonable to assume that these biases do exist to some degree. Researchers, editors, and reviewers —who are all drawn from the same pool of scientists, and who may tend toward having a strong personal conservation ethic—are perhaps likely to consider a result that supports climate change to be of greater interest than one that does not. The result of this could be a publishing bias that will skew our conclusions.

These are important issues, but let's take a brief look at a couple of important counterarguments. First, Camille Parmesan and Gary Yohe included in their meta-analysis an attempt to control for any publishing bias. For one of their tests, these researchers only included published studies that reported results for multiple species, some of which were changing in the direction expected due to climate change and some of which were not. By reporting both positive and negative results, these studies are the least susceptible to the accusation of bias. The number of studies included was inevitably reduced, but more than 250 species from a wide range of different life forms were incorporated in the analysis. And the result? Same as before: the reduced dataset showed a clear trend for upslope and poleward distribution shifts and advanced spring phenology.

A second counterargument concerns the nature of scientific publishing. One important ingredient for any study submitted for publication in a scientific journal is that it tells us something new. Without a touch of novelty, a manuscript is likely to be returned to its discouraged author with a rejection letter stating that, although the study appears to be scientifically sound, it doesn't tell us anything we don't already know. The upshot of this is that as more and more studies report results that show impacts of climate change, we should expect it to become increasingly difficult to publish similar results and easier to publish counter results. What could be more novel than a study that flies in the face of climate change! While I won't go so far as to claim that it would now be easier to publish studies that run counter to climate change, this argument does bring into question the opposite possibility that publishing greatly favors studies that are in agreement with a warming climate.

And so, we can conclude our search for fingerprints of climate change with *very high confidence* that climate change is already affecting living systems.[23] But don't get used to this level of confidence. As we begin to look to the future and try to predict impacts over the coming century, our confidence will begin to take a nosedive.

Attribution

An additional question I want to address is whether the biological changes we've seen can be attributed to warming that has been caused by human activity. It is one thing to say that plants and animals are responding to warming, but quite another thing to say that plants and animals are responding to *anthropogenic* (human-caused) warming. The approach that the IPCC takes in this regard is to break the issue into two separate questions: First, can changes in biological systems be attributed to climate change? And, second, can a significant proportion of climate change be attributed to human activities? The sum of these two parts constitutes what the IPCC calls "joint attribution."

We have seen in this chapter that the answer to the first question is a resounding yes. As for the second question, you will recall from Chapter 1 that the IPCC considers it *very likely* that warming since the mid-twentieth century has been caused largely by human emissions of greenhouse gases. It is therefore reasonable to combine these two lines of evidence to conclude that anthropogenic warming is indeed responsible for impacts on biological systems. As for the likelihood that this conclusion is correct, it is always the case (indeed, it is statistically inevitable) that confidence is reduced when requiring that two lines of argument, neither of which is 100 percent certain, fall into place. The IPCC therefore downgrades *very likely* to *likely* when stating the likelihood that anthropogenic climate change is impacting the planet's biological systems.

Now, you might be left wondering about the significance of the first part of this joint attribution argument: is it surprising to find that plants and animals are responding to climate change? Since climate and biodiversity are closely related (as emphasized in Chapter 1), it is, in fact, to be expected that changes in climate will be reflected in biological systems.

However, there are at least two key reasons why our observations of biological impacts of climate change are important: First, solid evidence of the impacts of climate change holds a great deal of weight when trying to convince policy makers to take note of the issue. Presentation of concrete examples and meta-analyses by influential organizations such as the IPCC (which shared the 2007 Nobel Peace Prize for its work) has had a huge influence on the politics of climate change. And, second, observed impacts from the twentieth century give us clues as to what we can expect to see in the twenty-first century. If the climate continues to warm, as is expected, then it is reasonable to assume that the trends we've seen in recent decades will continue into the future.

All the examples we've looked at so far provide the backdrop for the question we will begin to address in the following chapter: what is the risk that a significant proportion of species will go extinct due to climate change?

CHAPTER 6

Extinction Risk

Experimenting with the Karoo

An ability to forecast future events has been one of the major achievements of scientific endeavor, setting it apart from more hokum sources of knowledge ever since The Enlightenment of the seventeenth century. It is remarkable, for instance, that astronomers can apply Newton's laws of motion to predict with great certainty that Halley's Comet will next be visible from Earth in the year 2061; however, it remains a much greater challenge to forecast what the climate will be like around the time of this celestial wonder, or how biodiversity will have responded to the changing conditions by then. In this chapter, we begin to address this challenge by shifting our focus from observations of the past to predictions of the future.

One initial line of inquiry is to ask whether many species are already near the limits of their temperature tolerance—are organisms likely to survive temperatures much higher than today's? Global climate has, in fact, been generally cooling over the past 14 million years, and many species in existence today therefore arose under warmer conditions. It is therefore possible that many species may have retained tolerance for warmer conditions.

However, Earth's climate over the last 2 million years has been dominated by the dramatic temperature swings of glacial cycles, and it is unlikely that species accustomed to warmer climates from millions of years before would have survived these cold periods without becoming better adapted to cooler conditions. It is therefore plausible that the warm interglacial periods of the last 2 million years provide an appropriate benchmark for species' tolerance to high temperatures. Which leads us to a worrying fact: recent warming has driven the global mean temperature to within roughly 2°F of Earth's maximum temperature over the past million years. This means that, if projections of future climates are correct, species will soon be subject to conditions that are warmer than have been experienced for at least a million years. This does not bode well for species' survival.[1]

Attempting to estimate species' tolerances to warming by considering climates of the distant past is a useful starting point, but it is speculative to assume that organisms alive today maintain tolerance limits similar to those of their ancestors from millions of years ago. A more direct way to get a handle on temperature tolerances is through old-fashioned scientific experimentation: heat some organisms up and observe what happens. By artificially cranking up the temperature and studying how species respond, we can gauge how much warming can be tolerated before an organism overheats and dies. Before you get concerned about animal cruelty, I should clarify that this approach is mostly undertaken with plants. The example I'll describe below takes us back to the plants of the Succulent Karoo, which we last looked at in Chapter 3.

Experiments to test the tolerance of plants to warming have been undertaken at two remote sites near the small settlement of Vanrhynsdorp, roughly 300 kilometers north of Cape Town, South Africa. In late 2002, during the southern hemisphere spring, researchers from the South African National Biodiversity Institute set up 18 transparent plastic chambers around communities of small succulent plants at the two sites. The chambers were designed to act like mini-greenhouses, but were left open-topped to avoid excessive heating. The plants living within the chambers were inventoried, as were plants in "control plots" that were not subject to artificial warming. Each chamber and control plot was then fitted with an automatic data logger to record the local temperature, and the chambers were left alone—a rather alien addition to the landscape —for 4 months.[2]

On returning to the sites in mid-autumn, the researchers downloaded the temperature data and once more inventoried all sites, counting the number of dead plants and the fractions of dry, lifeless leaves in the canopies of those that survived. Maximum daily temperatures within the chambers were found to have been 9.9°F higher than at the control plots. This degree of warming is on the high side of what is predicted for this part of the world, but the warming does provide a rough approximation for "worst-case" twenty-first-century climate change. It is worth noting that the experimental warming occurred almost instantaneously—rather than increasing relatively slowly over the course of a century—but the experiment still tells us something useful about how much heat these plants can tolerate.

So how did the plant communities in the mini-greenhouses look when the researchers returned? Well, not so good. Populations of some species experienced mortality rates up to five times higher than those in the control plots, while other species showed they were under strain by significantly increasing the fraction of dead leaves on those plants that managed to stay alive. These experiments consequently reveal that a temperature rise in the order of 10°F is sufficient to exceed—or, at least, to nearly exceed—the lethal temperature limits of these succulents.

Traditional experimental techniques can therefore help us begin to get a handle on how biodiversity may react to future climate change. By subjecting organisms to a range of different warming scenarios (note that in the Karoo example, only a single scenario was tested) it is possible to find the temperature range within which a species can live. And if a sprinkler system is added to the experimental setup, it is also possible to find the range of precipitation within which a species has sufficient moisture to survive.

But beyond finding tolerance limits, researchers also use experiments to understand exactly *how* species respond to changes in temperature and precipitation. For example, we can assess how the moisture content of leaves changes, or how the chemical composition of the plant adjusts. Detailed, painstaking work such as this has been used to generate a *mechanistic*—or *process-based*—understanding of how plants respond to changes in climate. By formulating this understanding into equations that describe how plant physiology responds to temperature and moisture changes, researchers have been able to build sophisticated computer simulations called global vegetation models that estimate which types of plants will grow in different regions of the world under global climate change. Because these models require very detailed knowledge about plant physiology, it is more practical to model groups of species—termed *biomes*—rather than individual species, as I will explain next.

Shifting biomes

Biomes describe the major communities of species that are found in different climates around the world. They are defined based on factors such as life form (trees, shrubs and grasses), life-cycle approaches (deciduous versus evergreen), leaf types (broadleaf versus needleleaf), and density of plants (ranging from thick forest to sparse savanna and desert). The number of categories that should be identified is a point of debate, but it is common to divide the planet's major ecosystems into roughly 20 different types, including, for example, tropical rainforest, temperate conifer forest, temperate deciduous forest, moist savanna grassland, and dry desert.

Based on a mechanistic understanding of how the dominant types of plants in each biome function under different climatic conditions, global vegetation models predict which category is most likely to occur under a given set of conditions. Tolerance limits are assigned to each biome based on a sample of representative species for which responses to temperature and moisture are well understood. To illustrate this approach, we can consider the constraint that minimum temperature imposes on plant growth. For instance, broad-leaved evergreen trees are able to withstand occasional frost, but generally suffer severe leaf damage when temperatures drop below 5°F.

By contrast, broad-leaved deciduous trees are able to withstand much colder temperatures, commonly in the region of -40°F. One of the mechanisms by which deciduous trees survive very low temperatures is, of course, by shedding their leaves; basically, they shut up shop for the winter by not maintaining leaves that are susceptible to extreme cold, and that would be a drain on resources during the months when the potential for photosynthetic activity is low. Other species take a very different approach to surviving frigid conditions. Some spruce and fir trees survive temperatures as low as -76°F by expelling water from their tissues, creating a freeze-drying mechanism that preserves plant tissue much like the process of freeze-drying is used to preserve foods such as instant coffee.

By applying similar knowledge of plant responses to a range of different climatic factors—including minimum and maximum temperatures, length of the growing season, and moisture availability—global vegetation models work out which biomes could occur in each region of the globe. In cases where more than one type of biome could occur, the models select a single prediction according to a hierarchy that prioritizes more competitive types. For instance, tropical evergreen forests are given first priority, warm temperate evergreen forests are next in the hierarchy, and desert shrubs lie at the bottom of the pile. So according to this prioritization, tropical forests will dominate any area where climatic conditions are right for their growth (even if other biomes could potentially occur), whereas desert shrubs are restricted to only those marginal areas where no other plants compete. Using this approach, global vegetation models have proven successful in re-creating the present-day global distribution of biomes, which offers some hope that the models will be capable of predicting future distributions under an altered climate.[3]

Jay Malcolm, a professor at the University of Toronto, has applied these models to estimate the number of species that are likely to go extinct due to climate change. Malcolm and his team focused their analyses on 25 of the planet's biodiversity hotspots—a good place to start since these hotspots cover just 1.4 percent of the global land area, yet are home to roughly 44 percent of the world's plants and 35 percent of its vertebrate animals. We've already looked at some of these hotspots, including Madagascar, the Succulent Karoo in South Africa and Namibia, and Costa Rica (part of the broader Meso-american hotspot); other examples include the Californian Floristic Province, the Caribbean, the Mediterranean Basin, Southwest Australia, and the Guinean Forests of West Africa. Malcolm's strategy was to use global vegetation models to predict how climate change will alter the distributions of biomes within each hotspot, and to then estimate the number of endemic species that might be lost in cases where biome distributions shrink.[4]

A crucial element of this approach—and also some of the other approaches that I'll describe later in this chapter—is application of the "species–area relationship," which describes the number of species that are expected to be found in a given area of habitat. Based on field surveys, primarily on islands of different sizes, it has been shown that the number of species within any given area, and the size of that area, are related in a way that can be described by a mathematical formula.[5] We can avoid the symbols of mathematical notation here; basically, the number of species in an area decreases at an accelerating rate as the area becomes smaller. The upshot of this is that it is possible—within certain bounds of confidence—to estimate the number of species that you would expect to lose or gain as an area of habitat changes in size. The predictions may not provide the same level of accuracy with which Newton's laws can predict the motion of celestial bodies, but the species–area relationship is about as close to a law of nature as we have in ecological science.

Another important methodological issue that Malcolm and his team had to confront concerned the varying dispersal abilities of different species. Their solution was to assess the full range of possibilities by considering two extreme cases: First, they assumed no dispersal at all, meaning that biomes will only occupy parts of their current range that remain climatically suitable in the future. Second, they assumed unlimited dispersal, meaning that biomes will be able to shift across the landscape to occupy any area that becomes climatically suitable in the future. I've already given examples of species that are likely to fit either end of this spectrum; for example, extremely slow movers such as the Quiver tree are likely to adhere to the "no dispersal" assumption, while some highly mobile birds fit the "unlimited dispersal" model. In reality, the majority of species are likely to have dispersal abilities that fall somewhere in between these two extremes, but when predicting future impacts, it is sensible to consider the full range of possible outcomes.

Putting all this together—global vegetation models, the species–area relationship, and assumptions about dispersal ability—Malcolm and his colleagues had an approach that enabled the number of species at risk of extinction to be estimated. Their conclusions make for sobering reading. At the low end, if all the unknowns work in nature's favor, only one percent of species in biodiversity hotspots were predicted to be at risk of extinction. This is assuming, in particular, unlimited dispersal. But at the high end, projected extinctions amounted to roughly 40 percent of the biota, representing the potential loss of around 56,000 plant species and 3,700 vertebrate animals. Hotspots identified as especially vulnerable were South Africa's Cape Floristic Region, the Mediterranean Basin, and Tropical Andes, each of which were predicted to lose more than 3,000 endemic plants; and the Caribbean, Indo-Burma, and Southwest Australia, where more than 200 vertebrates were predicted to go extinct in each region. Overall, Malcolm's team predicted that climate change over the coming century could lead to the eventual loss of thousands, perhaps tens of thousands, of species from what are currently Earth's most biologically diverse regions.

An important point to emphasize is that there was a wide range of possible outcomes predicted by the models, from one percent of species going extinct at the low end to 40 percent extinct at the high end. Such a wide range of possible outcomes is typical of studies such as this and reflects the many unknown factors that must be accounted for. How far and how fast will species be able to disperse? How much will the climate change? Wherever possible, the general approach to dealing with unknown factors is to explore the full range of possibilities, just as Malcolm and his colleagues did by assuming either no dispersal or universal dispersal. But other unknown factors—such as the potential rate of evolutionary adaptation—remain nearly impossible to estimate for large numbers of species and are commonly ignored in current approaches. We will tackle in more detail some of the main factors that lead to uncertainty in predictions of extinction risk in the next two chapters.

First, I want to consider a very different line of thinking for generating estimates of extinction risk. One limitation of the global vegetation models that I've described is that they model groups of species—biomes—rather than individual species. A clear message from the fossil pollen record is that species respond to climate change individualistically: as the climate warmed at the end of the last glacial period, species shifted their ranges at very different rates, rather than moving in cohort. So we must question the wisdom of estimating species' extinction risk based on models of biomes—perhaps the groups of species that we observe today are merely temporary gatherings that will disband in the future.

An alternative, then, is to find a way of estimating future impacts on individual species. Scientists have developed a rather nifty methodology—based on defining what are termed *bioclimate envelopes*—that fits this bill. The approach has limitations of its own, but has been applied in a number of key influential studies.

Envelopes into the future

It is impractical to run experiments for a large number of species in order to understand the climatic conditions that they can tolerate. However, one thing we already know for many species is their present-day distribution. An alternative to running experiments is therefore to ask: What climates does each species currently occur in? For instance, Wendy Foden's grand tour of Namibia and South Africa resulted in a pretty good picture of where Quiver trees are currently distributed (even though some populations were found to be healthier than others; see pages 56–60). By studying the climate at sites where Quivers have been observed, it is possible to get an idea of the kinds of conditions that the species seems to like,

including upper and lower bounds of temperature and precipitation. This set of conditions defines the bioclimate envelope
of the species—it is the set of climate conditions found across
the range of the species.[6]

In practice, a species' bioclimate envelope is usually
identified using a computer, largely because some advanced
statistical methods come in handy when trying to characterize
the envelope based on multiple climate variables.[7] Instead of
simply working with upper and lower limits for each variable,
statistical methods can be used to estimate the probability
that a species will occur at intermediate values. For example,
we might find that a species of broad-leaved deciduous tree
is found within a range of temperatures spanning -40°F to
+50°F, but statistical methods may tell us that the species has
a particularly high probability of occurring between, say, 5°F
and 30°F, and a much lower probability toward its tolerance
limits. Moreover, advanced statistical methods can also look
for relations between variables: perhaps our tree can survive
temperatures down to -40°F, but only in situations where
precipitation is high and there is little wind.

The bioclimate envelope approach, then, is to create a
computer model that describes the types of climate that a
species lives in. Using this model, it is possible to draw on
a map those locations that are predicted to have conditions that
are favorable for the species under a scenario of future climate
change. A bioclimate envelope model therefore simply predicts where the climate conditions that the species currently
inhabits will be in the future. If suitable climate disappears in
the future, or shifts to an area that the species cannot reach,
then we can infer that the species is at risk of extinction
from climate change.

It is important to be clear how this approach differs from

the experimental methods we looked at earlier in this chapter. Experiments can be used to determine the tolerance limits above or below which an organism will die; by contrast, bioclimate envelopes only identify the range of conditions that a species experiences at the current time. In reality, a species may not experience the full range of climate conditions that it can tolerate. For instance, many species have their distributions curtailed by barriers such as mountains, rivers, or oceans, or by competition with other organisms, meaning that they may not inhabit areas that are toward the edge of their climatic tolerance. (I described a nice example of this in Chapter 3, where competition between barnacles was shown to restrict species to only a subset of the water temperatures they can tolerate.) This implies that bioclimate envelope models might not identify the full range of conditions under which a species can survive, meaning that the approach will tend to underestimate the true tolerance limits of species.

However, an alternative line of thinking suggests that bioclimate envelope models give a more realistic estimate of the range of values that species will actually be able to occupy in the future: just as species today do not experience the full range of conditions that they can tolerate, neither will they in the future. Predictions from bioclimate envelopes are therefore difficult to interpret, and are far from perfect, yet the approach gives us a different perspective on the extinction risk problem. Equipped with an understanding of how the models work, we can move forward to look at some of the conclusions that have been drawn.

Let's start by looking at predictions for a single species: what is the outlook for the Quiver tree under future climate change? Wendy Foden's team used bioclimate envelope modeling to predict that suitable conditions for Quivers will shift poleward by roughly 100 kilometers, and upslope by around 75 meters, over the next half century. These results are consistent with the trend that has been observed in the field, where populations at the cooler northern and low altitude ranges are suffering increased mortality. The model results add to our previous knowledge by quantifying how far the species would be required to move in order to shift its distribution and keep pace with the changing climate—the Quiver, it seems, would need to migrate poleward by about 20 kilometers per decade, which is mighty unlikely given that we know this long-lived species is not a fast mover. Once again, prospects for the Quiver look rather ominous.[8]

So we can get a taste of the future for an individual species, but a big advantage of the bioclimate envelope approach is that it can easily be applied to hundreds, or even thousands, of species, provided we have a fairly good idea of their current distributions. For an example of a study that modeled many species, we can shift our focus to the Fynbos biome, which dominates the cooler and wetter lands to the south of the Succulent Karoo in South Africa. Fynbos is one of the world's most diverse and distinctive floras, comprising more than 8,000 species of flowering plants, almost 70 percent of which are endemic to this region. Most well known among these plants are the proteas, whose large spectacular flowers are widely disseminated as part of the international flower trade. The King Protea, in particular, produces flowers in excess of 25 centimeters in diameter and is recognized as South Africa's national flower.

Given their economic and cultural importance, the proteas have been widely studied, especially through a major national initiative in the 1990s to map their distributions.[9] Thanks largely to the efforts of dedicated volunteers, more than 40,000 sample sites were assessed to give a remarkably complete picture of distributions throughout the region—information that is ideal for bioclimate envelope modeling. Soon after the data was compiled, a research team led by South African botanist Guy Midgley ran models for all 330 proteas that are endemic to the Fynbos biome, with the goal of drawing general, florawide conclusions as to the potential impact of climate change on the Fynbos biome.[10]

Once again, the results are alarming: for roughly one third of the species, no part of the current distribution is predicted to remain climatically suitable by the middle of the twenty-first century. These species may therefore need to shift their distributions elsewhere in order to fend off extinction, yet much like Quiver trees, proteas are notoriously slow movers. In fact, protea seeds are largely moved around the landscape by ants, which tend to disperse them just a few meters—certainly not the kilometers that would be required under climate change.

All in all, the view from South Africa seems fairly bleak. But let's now briefly look at some studies that modeled multiple species elsewhere in the world.

Kansas University biologist Town Peterson has led an analysis for a big chunk of the fauna of Mexico. Peterson and colleagues collated information on the current distribution of 334 species of birds, mammals, and butterflies that are endemic to Mexico, and then modeled the bioclimate envelope for each species using two mid-twenty-first-century climate change scenarios—one optimistic and the other pessimistic. The models predicted that half of the species studied will contract their distributions, even under the best-case scenario of unlimited dispersal and less severe climate change. As for the worst-case scenario (assuming no dispersal and more extreme climate change), all species suffered contractions in their distributions, with roughly one in five potentially losing more than half of its range. So the picture from Mexico isn't too bright, either, but it is important to note that only 2 percent of species were predicted to contract their range by more than 90 percent, which would represent an especially grave danger of extinction.[11]

What about Europe? French researcher Wilfried Thuiller has led a study that modeled a whopping 1,350 plant species, representing around 10 percent of the European flora.[12] Thuiller's study used the usual assortment of climate and dispersal scenarios, and concluded that under a worst-case scenario, up to 22 percent of species could lose more than four-fifths of their range within a century. By contrast, at the less severe end of the range of predictions, only 2 percent of species were predicted to lose four-fifths of their range. But before we allow a glimmer of optimism . . .

How about Australia? Stephen Williams and his colleagues at James Cook University focused their attention on the wet tropical rainforests of eastern Australia, which are home to 65 endemic vertebrate species, including a number of uniquely Australasian possums and tree kangaroos. Under a temperature rise of 6.3°F—which is within the bounds of what has been predicted in this part of the world for the twenty-first century—they found that bioclimate envelopes for nearly half of these endemics will disappear, creating what Williams has dubbed "an impending environmental catastrophe."[13]

It seems that whichever way we look at it—using global vegetation models or bioclimate envelopes, or analyzing biomes versus individual species—the finding is that climate change has the potential to cause masses of extinctions around the world. There are considerable differences between the best-case and worst-case scenarios that we've looked at, but it is inevitable that the more extreme scenarios of the future will capture most public attention. With this in mind, I will close this chapter by examining in detail the most high-profile and dramatic predictions of global extinction risk from climate change that have been made to date.

Committed to extinction

If you want people to take notice of your research—as has most certainly been the case with the study we're about to look at—then it pays to think big. In 2004, two years before Jay Malcolm published his global analysis of hotspots, Chris Thomas led the first big global assessment of extinction risk from climate change.[14] Thomas, some of whose work we encountered in Chapter 3, was able to pull together a large group of collaborators from around the world. The group pooled their data and, using an approach not dissimilar to the meta-analyses described in Chapter 5, undertook a combined analysis of results from multiple individual studies. In total, they collated data from sample regions that cover approximately 20 percent of the planet's terrestrial surface, including three of the regions we've already looked at—South Africa, Mexico, and Australia—as well as two additional ones: the tropical forests of Amazonia and Brazil's tropical savanna (known as the cerrado).

Thomas's study not only included a diverse set of regions, but also a diversity of species, including mammals, birds, amphibians, reptiles, butterflies, and plants. In all, the research team assessed 1,103 species that are endemic to the study regions. Each species was modeled individually using the bioclimate envelope approach, and the results were analyzed under the usual two assumptions of either zero dispersal or unlimited dispersal. As for scenarios of global climate change, the research team chose to examine projections that extended until the middle of the twenty-first century and represented three levels of severity: minimum expected change 1.4–3.1°F, midrange change 3.2–3.6°F, and maximum expected change 3.7–5.4°F.[15]

The number of species that might become extinct was estimated using two methods: first, applying the species–area relationship, which we will return to shortly; and second, using criteria established by the International Union for Conservation of Nature (IUCN) for assigning species to its "Red List." The IUCN Red List is widely regarded as the world's most authoritative inventory of threatened species—a kind of "who's who" of extinction risk—and is commonly used to target conservation policy and spending. Most species on the Red List are there because their existence is threatened by factors including habitat destruction, hunting, or competition from newly introduced species, but Thomas and his colleagues realized that similar criteria for assigning risk from these threats might also be applicable to climate change.

The criteria used by the IUCN represent the best available knowledge of the extinction risk associated with a species' population falling below a certain size, or its range shrinking below a certain area. A considerable advantage over methods we've looked at so far is that specific probabilities of extinction are assigned to particular circumstances. Instead of concluding that a certain reduction in range conveys a "high" risk of extinction, the Red List criteria specify a probability—say, 3 out of 4—that the species will die out. Based on the IUCN's approach, Thomas and his collaborators assigned species to four levels of threat. For instance, species that were predicted to experience a contraction in range by more than 80 percent, or whose range was predicted to shrink to less than 10 square kilometers, were classed as *critically endangered* and assigned a probability of extinction of 3 out of 4. Following similar rules, other species were categorized as either *endangered*, carrying an extinction risk of 1 in 3, or *vulnerable*, carrying a risk of 1 in 6. Species whose range was predicted to shrink to nothing were, not surprisingly, classified as *extinct*. No probabilities needed in that case: if you no longer have anywhere to live, then your number's up.

So, by tallying the number of species that fell into different threat categories, it was possible to estimate the proportion of species that might become extinct due to climate change. Once again, there was a substantial spread in results across different scenarios, with a minimum of 11 percent of species predicted to go extinct (assuming minimum climate change and universal dispersal) and a maximum of 58 percent (maximum climate change and no dispersal). If the results are averaged across the two dispersal assumptions, then roughly 23 percent extinction is predicted for minimum climate change, 32 percent for midrange climate change, and 46 percent for maximum climate change. According to this view of what the future may hold, nearly one quarter of the species studied by Thomas and his colleagues will go extinct even if climate change is more toward the best-case end of what is expected.

But perhaps the IUCN Red List approach is too pessimistic—what if the species–area relationship is used instead? Thomas's group applied the relationship to the contractions in range predicted by their bioclimate envelope models.[16] (This differs from the use of the species–area relationship by Jay Malcolm earlier in this chapter, where range contractions were predicted by global vegetation models.) If the results are again averaged over the two dispersal scenarios, 18 percent of species were predicted to go extinct under minimum climate change, 24 percent under midrange warming, and 35 percent under maximum climate change. So the results from the species–area approach are no less worrying.

It is important to clarify that neither of the two approaches we've looked at for estimating extinction risk from range contractions—the species–area relationship and the IUCN Red List criteria—are intended to predict exactly when extinction will occur. The approaches are based on predicted reductions in the area of land that is climatically suitable, but the possible time lag between loss of suitable area and extinction remains poorly understood; in reality, decades may elapse before the predicted extinctions take place. Therefore, although Thomas and colleagues' predictions are for climate change around the year 2050, it is not expected that all predicted extinctions will have occurred by that time. The authors are careful to refer to their estimates as being proportions of species that are "committed to extinction." It is kind of like a death sentence without a specified execution date.

In sum, Thomas and colleagues' study predicted that the diversity of life on Earth will be severely impoverished by twenty-first-century climate change. The story buzzed along media newswires and it has been estimated that, in the United States alone, the total readership of newspapers covering the story was greater than 21 million, while more than 13 million people saw television news items on the subject. The article was also presented before the U.S. Senate, discussed in the UK House of Commons, and led to a number of statements by senior politicians. Furthermore, the story has been widely debated on the Internet, with many of the best-known international conservation organizations, including the World Wildlife Fund (WWF) and Greenpeace, running commentaries on their websites.[17] The Thomas study was in this way responsible for substantially raising awareness of the issue.

We'll pick up this story again in Chapter 9, where we'll address whether the risks associated with climate change are being oversold—are headlines forecasting masses of extinctions justified, or are they alarmist? To address this question fully, it is necessary to first delve deeper into some of the uncertainties surrounding the predictions of extinction risk that we've looked at in this chapter. So, as promised, in the next two chapters we will take a close look at two key factors—evolutionary adaptation and community impacts—that remain poorly understood and yet could have a very substantial impact on the accuracy of current predictions of extinction risk.

CHAPTER 7

Running to Keep Still

A Darwinian perspective

In the topsy-turvy world of Lewis Carroll's *Through the Looking Glass,* the Red Queen explained to a puzzled Alice that "it takes all the running you can do, to keep in the same place." The Red Queen, it turns out, never gets very far because the landscape moves with her as she runs! This predicament faced by the Red Queen is often quoted by evolutionary biologists as being analogous to the situation faced by species, which are required to continually evolve in order to maintain fitness relative to the changing world around them. Most commonly, the analogy is used to refer to species that are entrenched in an evolutionary arms race with one another, but the idea is also relevant for our interests: as climatic conditions change, it takes all the evolving a species can do to keep in the same place.

The alarming predictions of massive species loss that we

looked at in the previous chapter were generated based on the assumption that species have two options when faced with rapid climate change: either they shift their distribution to find more favorable conditions, or they go extinct. But another possibility is that species will be able to adapt *in situ* to cope with the new climate conditions. Perhaps many species will undergo rapid evolutionary selection—survival of the fittest— in favor of individuals that are best adapted to warmer conditions, resulting in a shift in the range of conditions that can be tolerated. This process might provide an important buffer against extinction, meaning that many of the most catastrophic predictions of mass extinction are overly pessimistic.

It is usually expected that evolutionary change occurs only on long time scales; indeed, Charles Darwin emphasized that vast periods of geologic time are required to allow the accumulation of slight differences from one generation to the next, which eventually result in major differences in biological form. It is often thought, therefore, that the range of temperatures that a species can tolerate will not change over time frames that are relevant for rapid climate change. However, although the evolution of major new forms—new species—may take millions of years, we will see in this chapter that smaller, more subtle adaptations can accrue within a remarkably short time frame.

One way we can begin to get a handle on the potential for rapid evolution is through "common garden experiments," in which plants are transferred from one location to another where the climate is different. Let's start with the example of two perennial plants from the Asteraceae family—*Solidago altissima* and *Solidago gigantea*—which have been intensively studied using such experiments.[1] The bright yellow flower heads of these two species are a common sight throughout

Europe, but both plants are, in fact, native to North America and were only introduced to Europe in the eighteenth century —S. *altissima* in 1738, and S. *gigantea* in 1758, to be precise. From their initial introduction as ornaments in London, the plants were further redistributed throughout Europe and have since escaped the bounds of managed gardens and nurseries to establish natural populations in the countryside. Today, populations of both species flourish in a diverse array of climates across the continent, from the warmth of southern France to the chills of Scandinavia.[2]

So each of the species appears to be tolerant of a wide range of different climates, but common garden experiments reveal something interesting: when plants are transferred from their home environment to a different part of the species' range, growth tends to be much less vigorous at sites that are either warmer or colder than where they were transplanted from. What we find, then, is that populations throughout the species' range are adapted to local conditions. Although all individuals from France to Scandinavia are members of the same species—they can all fertilize one another—the range of climates that each can tolerate varies considerably.

So, what does this tell us about the potential for rapid evolution? Since there is no evidence that these Asteraceae were introduced into Europe multiple times, it seems that the present-day variation in climatic preference among populations has evolved from the initial London stock that was introduced in the eighteenth century. In that case, the speed of adaptation is impressive, with different responses to climate having accrued in fewer than three centuries. Clearly, populations *do* adapt.

But we're still talking about adaptation over a matter of centuries, yet climate change is usually discussed within a time frame of decades. For an example of even more speedy evolution, we can turn to a common garden experiment carried out in Britain on navelwort—a small perennial plant that is somewhat less showy than the Asteraceae, but can be recognized by the navel-like depression in its leaves from which its common name (navelwort) and scientific name (*Umbilicus*) derive. The natural range of navelwort extends through the west and south of Britain, and it is commonly found growing in damp rock crevices, such as along the stone walls that are typical of the English countryside.

In 1978, British botanist Ian Woodward collected seeds from a population of navelwort in Cardiff, Wales, which is within the species' natural range. Woodward transplanted these seeds roughly 250 kilometers east and 50 kilometers south to the town of Crowborough, where he attempted to establish a new population by spreading seeds on a stone wall.[3] Crowborough is outside the natural range of navelwort and has a climate that is generally cooler than that of Cardiff, which enabled Woodward to test the impact of cold temperatures on plant survival. By 1987, a healthy new population had established itself, from which Woodward collected seeds and plants for comparison with the "donor" population in Cardiff.

Samples from both the donor and transplanted populations were subject to experimental tests in order to determine whether responses to temperature differed. Seeds were planted in greenhouses at a range of different temperatures, while adult plants from each population were placed in a freezer, subjected to sub-zero temperatures for an hour, and then removed and given the opportunity to recover at more usual temperatures. Of course, these experiments test the navelwort's response to cooler, rather than warmer, temperatures—not what we're expecting under climate change—but the species' capacity to adapt to temperature change is relevant regardless of the direction of that change.

Ten years is a long time for a scientist running an experiment, but it is the blink of an eye in evolutionary time frames. What, then, could Woodward conclude from his efforts? The donor and transplanted populations were, in fact, surprisingly different in their responses to temperature. Seeds from the transplanted population had an enhanced capacity to germinate at low temperatures, while adult plants from the transplanted population were more likely to revive when rescued from the freezer. Overall, navelwort had managed to evolve new responses to low temperature, with temperatures that were observed to kill individuals from the Cardiff population endured by around half of the new settlers in Crowborough. So we can conclude that species may be able to adapt to new temperatures remarkable quickly—in the case of navelwort, over just a single decade.

Rapid evolution in plants is made possible by the fact that only a small fraction of seeds germinate and survive to adulthood. When seeds are transferred from one locality to another—either by natural dispersal or, in the case of navelwort, in the trunk of a researcher's car—variation in their genetic makeup means that only those best suited to flourish in the new environment will germinate and eventually pass on their genes to the next generation. For the navelwort, 400 or so seeds were moved, yet only a few of those—the ones most tolerant of the cooler conditions in Crowborough—survived to establish the new population.

In the examples we've looked at so far, for Asteraceae and navelwort, the selection process that drives evolution was imposed by moving species to novel conditions beyond their normal geographic range. Although our main interest is in how species will respond when the climate changes around them, we have begun to get a handle on the potential for rapid evolution by using climate change over space as a proxy for climate change over time. We can expect that a temporal change in climate will impose a similar selection pressure to a spatial change in climate, but opportunities to directly examine changes in temperature tolerance over time are very rare. However, in the next section, we will see how, under exceptional circumstances, seeds preserved for hundreds of years can be germinated and compared to their present-day ancestors from the same locality. Here we enter the rather astonishing field that has been dubbed "resurrection ecology."[4]

Resurrection ecology

Plant seeds can be extremely hardy little things. It is well documented that some seeds from herbaria and museums have remained viable for around 200 years and, occasionally, seeds much older than that have been revived. In one instance, nuts forming part of a necklace collected from a 600-year-old tomb in Argentina were successfully germinated,[5] and, in another case, 2,000-year-old date palm seeds excavated during archaeological digs in Israel produced healthy seedlings.[6] Such longevity provides a unique opportunity to resurrect and study populations of plants whose closest ancestors grew centuries ago.

An especially good place to start looking for very old, viable seeds is in the frozen soil of the Arctic tundra. As Arctic soil builds up, it buries seeds and, over time, creates a frozen "time transect" of seeds of varying ages. These seeds can be germinated and the resulting plants compared using experiments similar to those discussed earlier in this chapter for plants from different regions. Our main question, then, is whether individuals grown from old versus young seeds have different responses to climate.

Professor James McGraw and two of his graduate students at West Virginia University set about answering this question back in the summer of 1986.[7] The team selected two field sites in southern Alaska, braved the cold, and went digging for seeds. They excavated down to a depth of around 40 centimeters and brought up seeds predominantly from a type of sedge, scientifically named *Carex bigelowii*. To find whether any seeds were viable was a simple matter of planting them in a greenhouse and waiting—hoping—for germination. Fortunately, the efforts paid off and a total of 51 seedlings emerged.

These seedlings were gently extracted from the soil and the coating around the seed from which they had sprouted was carefully removed under a microscope. The seedlings were then repotted for later experiments, while the seed coats provided valuable organic material that could be dated to give the age of the seeds. The age of any dead organic material can be estimated using radiocarbon dating, which relies on a radioactive form of carbon called carbon-14. In short, when plants convert carbon dioxide into organic material during photosynthesis, they take in a minute amount of carbon-14, which occurs naturally in the atmosphere (albeit at very low concentrations, roughly one part per trillion of carbon). After plants die, radioactive decay causes the quantity of carbon-14 in their tissues to decline at a known fixed rate, so the concentration of remaining carbon-14 can be used to reliably estimate time since death. The ages of the seeds were dated using this method and found to be at least 200, and possibly nearer 300, years old.

After being repotted, the seedlings were left to mature for 6 months before clones of the plants were created by taking cuttings and rooting them, just as is commonly done in household gardening. In this way, populations of genetically identical plants were created, with each individual being the clone of a seedling that emerged from an old, dug-up seed. These individuals were then grown in experimental chambers alongside clones generated in an identical way from modern-day seeds collected at the same Arctic localities. To test for differences between the old and young stock, plants were subjected to three different temperature levels: 55°F, 72°F, and 81°F. The intermediate of these levels was chosen to roughly reflect the present-day average temperature during the growing season at the field site in Alaska, meaning that the experimental treatments on either side represented cooling and warming.

After 14 weeks, each plant was removed from its chamber and subjected to a rigorous health check: the numbers of leaves and stems were counted, the length of the longest leaf was measured, and—in a final act that destroyed the plant—the specimen was dried and weighed. So how did the populations from old stock measure up against their younger relatives? Well, that depends on the temperature regime to which they were subjected. In the cooler chamber, the older stock generally faired better: individuals that were genetically identical to plants from two centuries ago tended to achieve more body mass than their younger relatives, as reflected in higher dried weights. Moving to the warmer environments, both old and young populations responded positively, increasing their numbers of leaves and stems, and increasing total dried mass. But it was the modern-day populations that took most advantage of the warmer conditions, turning the table on their older relatives by achieving substantially more body mass at 81°F.

We have evidence, then, for different responses to temperature between the old and young populations. Temperature preference appears to have evolved in situ over the course of 200 years. Because the older populations fared better than the younger in cooler conditions, it seems reasonable to infer that recent climate warming may have had a hand in influencing these changes. Indeed, average temperatures in the Arctic have increased at almost twice the average global rate in the past century,[8] which will have imposed a selection pressure that is conducive for rapid evolution.

But, as always, we should not jump to conclusions before first considering alternative possibilities. One potential issue is the effect that centuries of lying dormant might have on seeds: is it possible that genetic damage occurred while the old seeds were buried and frozen? Breakdown of the chemical structure of DNA with age is indeed known to occur in seeds of some species, and McGraw's research team found a small number of albino seedlings emerging from old seeds, suggesting that genetic damage may have taken place. However, although genetic damage remains a possibility, it is unlikely that damaged seeds would have produced the healthy adult plants that were observed; in fact, plants from old seeds grew more vigorously than modern-day populations under cooler conditions.

Another, more plausible, explanation might be that seeds with a particular genetic makeup are better equipped to survive years of dormancy in frozen soil. Suppose that the seeds most likely to survive frozen dormancy are also the seeds that will produce cool-hardy plants. If that is the case, the individuals that McGraw and his team cloned from old seeds will not be representative of the population that existed in the past. Thus, our conclusion that old populations grew better under cool conditions would be mistaken—we would be fooled by a bias in the data.

So have we been fooled? Frankly, we can't quite be sure. Nobody has yet tested whether some seeds of *Carex bigelowii* survive better than others when buried for 200 years. But I'm not holding my breath in anticipation of this particular experiment being completed. We once again reach the edge of our knowledge and must draw conclusions based on our best understanding.

As we've seen earlier in the book, strength in an argument comes when multiple pieces of evidence from diverse sources all point in the same direction. So far in this chapter, we have seen how Asteraceae rapidly adapted to different climates across Europe, how a population of navelwort evolved to match conditions outside its normal range, and now—apparently— how an Arctic population two centuries ago was better adapted to cooler temperatures than its modern-day relatives. Taken together, these examples provide strong evidence against the notion that rapid climate change will inevitably overwhelm evolutionary processes.[9] Perhaps the Red Queen will be able to keep pace as the landscape changes around her.

However, the examples so far in this chapter have been exclusively for plants. Is rapid evolution also possible in the animal kingdom?

Adapting photoperiod

In Chapter 4 we looked at how climate change was affecting species' phenology, particularly with regard to the earlier arrival of spring. In each of the examples we considered, changes in life-cycle timing had occurred because individuals were following the same old rules, but under different circumstances. For instance, flowers on the Leopold family farm in Wisconsin were using the same temperature cues that they have always used for timing their blooms, but these cues were arriving earlier in the year. There was no need for us to infer evolutionary adaptation to explain the observed changes in phenology; rather, species were simply carrying on as normal.

An important finding, however, was that some species are not shifting their phenology in response to the changing climate. I concluded that in many cases this was because events are controlled by photoperiod (the length of daylight) rather than by climate. The result, of course, is a lack of synchronization between life-cycle events that are cued by climate and those that are cued by photoperiod. But we will now explore how rapid evolution may, in fact, enable some species to adjust their response to photoperiod under a changing climate.

For more than thirty years, biologists Bill Bradshaw and Christina Holzapfel have together studied the photoperiod response of the Pitcher-Plant Mosquito—*Wyeomyia smithii*—which is common in boggy habitats along the eastern seaboard of North America, from the Gulf of Mexico to northern Canada. This small mosquito has a remarkable eight-week life cycle, with its pre-adult development occurring entirely within the water-filled leaves of a carnivorous plant—the Purple Pitcher Plant. The "pitcher" from which the plant gets its name is a rainwater-collecting receptacle formed from fused leaves. Lured by the plant's nectar, hapless bugs fall into the pitcher and drown, providing a steady supply of nutrients. But since mosquitoes spend much of their life in water, Pitcher-Plant Mosquitoes don't fall prey to the trap; instead, they take advantage of it by laying their eggs directly on the pooled rainwater within the pitcher plant. The trap not only provides a sheltered, cozy environment for the development of mosquito larvae, but also a constant supply of food in the form of rotting carcasses.

During the fall, the Pitcher-Plant Mosquito life cycle grinds to a halt as larvae enter hibernal diapause—a dormant state similar to that of the Winter Moth that we saw in Chapter 4. As winter passes by, the mosquito larvae remain frozen in time, often crowded into a small pocket of water and nutrients covered by a block of ice. Then, during the spring and summer, the larvae pupate and adults emerge to perpetuate the mosquitoes' life cycle. This, of course, is the time of year when campers get the bug repellent out.[10]

It turns out that the timing of onset and termination of diapause in these mosquitoes is mediated by photoperiod. Bradshaw and Holzapfel can demonstrate this in their lab at the University of Oregon: the mosquitoes can be tricked into beginning or ending diapause by using artificial light to mimic different day lengths. By contrast, adjusting the temperature doesn't work. So, in the wild these mosquitoes take their cue to enter diapause from the shortening days of late summer, and to emerge from diapause from the lengthening days of spring. We therefore wouldn't expect that climate change will cause shifts in phenology for this species. However, Bradshaw and Holzapfel have discovered—quite by accident—that this is not, in fact, the case.

The evidence dates back to the early 1970s, when Bradshaw

and Holzapfel started collecting Pitcher-Plant Mosquitoes and transporting them back to the lab for experiments. In the early days, their research agenda had little to do with climate change, but as collecting continued through the subsequent decades, it became clear that something curious was occurring. Experiments revealed that the critical photoperiod at which the mosquitoes were entering diapause was getting shorter. In more recent years, the mosquitoes were waiting longer before retiring for the winter, such that by 1996, Pitcher-Plant Mosquitoes were entering diapause nine days later than in 1972. It seems that they are managing to take advantage of the longer summers (and hence, longer growing seasons) that recent climate change is offering, despite the fact that their phenology is not directly related to climate.[11]

Unlike in the previous examples of shifts in phenology that we've looked at, we now need to explain a shift in species' behavior; this species is no longer responding to the environment in the same way it did a few decades ago. By studying mosquito genetics, Bradshaw and Holzapfel have pinpointed the evolutionary process by which this has occurred. Because of genetic variation in a population, individuals enter diapause at slightly different times, with some reacting to shorter photoperiods and therefore entering later. As the climate has warmed and growing seasons have lengthened, more of the late-hibernating individuals have survived the winter because they minimize time spent in diapause, during which limited nutritional reserves are at risk of becoming depleted. The trait of entering diapause later is then passed on to the next generation such that, as the process is repeated over multiple years, most of the population is entering diapause at the time that is optimum to ensure survival through the winter.

Evolution thereby provides a crucial way for Pitcher-Plant

Mosquitoes to stay in tune with the changing environment. Although this species' phenology is not controlled directly by climate, it has nonetheless adapted its behavior so as to maximize its survival as seasonal patterns have changed. And, once again, this evolution has been speedy, with changes occurring in a time frame of decades. I don't know of any observations documenting changes in the phenology of the mosquitoes' host plant, the Purple Pitcher Plant, but the rate of advance in the mosquitoes' spring phenology—roughly 3½ days per decade—is in the same ballpark as rates we saw in Chapter 4 for some North American plants. So it is reasonable to conclude, in this case at least, that synchronization between phenological events controlled by climate and by photoperiod will not necessarily be severely disrupted by climate change.

Nature is proving more robust than we might otherwise have thought, keeping a number of nifty tricks up its sleeve to cope with environmental change. In the following section, we'll look at another such trick: the ability to quickly evolve improved dispersal capacity, facilitating more rapid range shifts as the climate changes.

Cricket morphs

It is not uncommon for a species to exist in more than one body form, or *morph*. Think of jaguars, for instance, which are most commonly golden with black spots, but also come in a dark, silky black form. Or similarly, the Two-Spotted Ladybug is typically red with black spots, but can also be black with red spots. In such cases, two or more distinctly different morphs exist in a single, interbreeding population, and there are no intermediate forms—you are either a golden jaguar with black spots or you are a black jaguar, and never anything in between. The formal term for this phenomenon is *polymorphism*.

Some species of bush cricket have polymorphisms in their wing forms, including the Long-Winged Conehead (*Cono-cephalus discolor*), which comes both in long-winged and extra-long-winged forms, and Roesel's Bush-Cricket (*Metrioptera roeselii*), which comes in a short-winged form that cannot fly and a long-winged form that can. A study of these crickets in Britain (once again masterminded by Chris Thomas) has revealed a shift in their distribution over recent decades, the nature of which will come as no surprise: both species have expanded their distributions northward (poleward).[12] But a twist to the familiar story is that recently established populations were found to have much higher frequencies of the longer-winged morphs. In older, more southerly populations, the longer-winged forms tended to be very much in the minority, making up just a few percent of the population. But in newer, northerly populations, the number of longer-winged individuals was substantially higher, in some cases comprising upwards of 60 percent of the population.

This swing in morph frequency can be explained by a straightforward selection process: once climate change begins to drive a shift in distribution, those individuals that are best equipped to disperse are likely to be favored, meaning that founding populations comprise a relatively high proportion of good dispersers. In effect, climate change demands that species up and move, and natural selection responds by favoring those individuals most capable of doing so. And because longer-winged individuals are more likely to have offspring that are also longer winged, we can expect an overall enhancement of dispersal capacity in populations that are on the move. By contrast, in populations that have been established for some time and are more settled, less dispersive forms will become more prevalent because longer wings are a hindrance for a sedentary lifestyle and will therefore be selected against.[13]

The implications of this for the rate at which species are able to shift their ranges may be profound. The newly established populations are better equipped to cross barriers—roads, parking lots, backyards, agricultural fields—that would have represented major obstacles before warming started. As a result, it is estimated that the swing toward longer-winged individuals in these crickets has enhanced the rate of range expansion by as much as fifteenfold.

Development of improved dispersal ability may not be restricted to British crickets. In Canada, inflated frequencies of longer-winged morphs have been observed in invading ground beetles, while butterflies expanding their ranges across Europe seem to have evolved larger thoraxes, meaning bigger muscles for flight. And among plants, there is evidence from North America that Lodgepole Pine evolved more dispersive seeds during warming at the end of the last glacial period.[14] Together, these examples demonstrate that dispersal ability is not necessarily fixed—it may be substantially enhanced in the face of climate change. Once again, Nature is proving more equipped to deal with change than we might previously have suspected.

Salamander morphs

Natural selection for improved dispersal ability occurs as a species is shifting its range, but as a final illustration of how species are adapting to climate change, let's now return to where we started the chapter: will species be able to adapt in situ as the environment changes around them?

The critter of interest this time is the Eastern Red-Backed Salamander—an amphibian that is abundant in the leaf litter of forests in eastern North America. Population crashes among salamanders are very much part of the global amphibian crisis described in Chapter 2,[15] so any hint of rapid adaptation to cope with environmental change is worth taking a look at. The Eastern Red-Backed Salamander is, in fact, not always red backed—it comes in two primary morphs, one completely black and the other with a red, or occasionally yellow, stripe down its back. The two primary morphs live together and readily mate with one another, but as with the British crickets, frequencies of different morphs vary in different parts of the range. Studies of this spatial variation show clearly that the black morph is more associated with warmer areas, and the striped morph is more associated with cooler areas.

It seems there is a very good physiological reason for this difference in temperature preference between the morphs. Lab tests have shown that black individuals expend less energy when at rest (they have a lower metabolic rate), meaning that they are able to remain more active in warmer conditions. So the black morph is better adapted to warm conditions, and by similar argument, the striped morph is better adapted to cool conditions (its higher metabolic rate makes it relatively more active at lower temperatures). Thus, we have an instance where different morphs are more and less tolerant of different temperatures.[16]

So what should we expect to happen under climate warming? Striped individuals should decline at the expense of black ones. Researchers James Gibbs and Nancy Karraker have confirmed exactly this by compiling observations of morph identities for more than 50,000 individual salamanders dating back to the early 1900s.[17] During the twentieth century, climate warmed by around 1.3°F throughout the species' range and, accordingly, the probability of an individual being striped declined from 80 percent to 74 percent. This isn't a huge change—an adjustment of just 6 percent—but it represents a significant, long-term trend indicating morphological change in response to changing conditions.

It is important to note that global warming is unlikely to be acting alone in this instance. The full story probably also involves human exploitation of forests, notably for timber extraction, which tends to leave thinner canopies that let in more of the sun's rays. As a consequence of this degradation, the forest floor habitat that is home to Eastern Red-Backed Salamanders has generally become less dark, moist, and cool than they have previously been accustomed to. This localized change in climate (*microclimate*) has more than likely contributed along with global warming to cause the observed morphological changes. But, nevertheless, the important point for us here is that change in the environment—be it due to global or local factors—has resulted in an adaptive response, this time in a morphological trait that is directly related to temperature tolerance.

Who will adapt?

I have now described multiple examples that prove the possibility of rapid evolution in response to a changing environment. Remember that we are not talking about major adaptations in form, such as development of the elephant's trunk or the giraffe's neck, but smaller changes affecting temperature tolerance or photoperiod response. This is in stark contrast to the common perception that Darwinian evolution only occurs over millions of years. (Though, to be fair to Darwin, he might not have been at all surprised by the examples we've looked at in this chapter; in fact, he spent a great deal of time breeding pigeons and was well aware that some characteristics, including coloration and beak shape, could change rapidly when selected for under domestication.)[18] And so, adaptation is likely to lessen the impacts of climate change in at least some instances. Some of the catastrophic predictions of massive species loss that we looked at in the previous chapter, which do not allow for the possibility of adaptation, are therefore expected to overestimate, at least to some degree, the risks of extinction from climate change.[19]

But how common is adaptation to climate change likely to be? Is the potential for rapid adaptation restricted to a few unusual cases, or is it likely to be widespread throughout biodiversity?

You'll notice that I haven't described any instances of rapid evolution in long-lived species—say, grizzly bears, tigers, or Quiver trees. Instead, each example has involved species with a relatively short generation time, meaning that individuals reproduce relatively quickly, resulting in a rapid turnover from one generation to the next. Our most extreme example in this regard was the Pitcher-Plant Mosquito, which completes its entire life cycle—from egg to death—in just eight weeks. Since selection from one generation to the next is the engine of evolution, a shorter time span between generations will enable more rapid evolution. So among plants, there is a huge difference in the potential for rapid evolution between annuals, which live for just one season, and long-lived giants such as oak trees, which can live for centuries and rarely start producing acorns until they are at least 20 years old. Likewise for animals, we cannot expect many large charismatic vertebrates—pandas, elephants, bears, whales— to evolve in response to rapid climate change.

We can liken generation time to the speed at which the Red Queen takes her steps: the quicker she takes each step, the more chance she has of keeping pace with the changing landscape. The other crucial factor that will determine how fast she runs is the distance taken in each step: to move as fast as possible, she should take giant strides rather than fairy steps. In evolutionary terms, step size relates to how much adaptation takes place in each generation. The easiest way to think about this is with reference to plants. Most plants produce copious numbers of seeds—think of the hundreds of acorns produced each year by a single oak tree, or the number of fiery seeds you get inside each chili pepper—yet only a miniscule fraction of these seeds survive to reach maturity and reproduce. This reproductive strategy gives natural selection a large number of individuals from which to pick, providing the opportunity for big evolutionary steps. Contrast this with the few babies produced by most large mammals and we again see that rapid evolution is not equally likely among all species.

Another important factor controlling the size of each evolutionary step is the amount of genetic variation in a population. Because variation between offspring is a necessary ingredient for natural selection, it is no use producing thousands of progeny if they are all genetically very similar. In general, we expect larger populations to incorporate more diversity, yet many species nowadays have very small populations, restricted to fragmented scraps of habitat in a human-dominated landscape. Species that are affected by habitat fragmentation are therefore less likely to muster the genetic resources for rapid adaptation to climate change.

So we can conclude that long generation times, small numbers of progeny (seeds in plants, or babies in animals), and small population sizes all severely limit the potential for rapid evolution. Overall, the evidence we have does not suggest that the majority of species will be able to adapt at a sufficient pace to keep up with the changes in climate that are projected for the coming decades. Indeed, the numerous case studies we looked at in Chapters 2 through 5 provide evidence that rapid evolution isn't the norm. In cases where species' ranges are shifting upslope or poleward, populations at the trailing edge (those at lower elevations or toward the equator) are commonly dying out—they are not adapting to cope with new conditions, but are instead becoming locally extinct.[20] Likewise, numerous species whose phenology is thought to be controlled by photoperiod have not shown the same rapid adaptation as has been seen in Pitcher-Plant Mosquitoes. And in the case of corals, the widespread bleaching that has occurred suggests that populations are not rapidly adapting their genetic makeup to increase tolerance to warmer waters (probably because coral generation times are long, sometimes upwards of 30 years).[21] The instances we've looked at in this chapter are therefore more likely to represent the exceptions rather than the rule.

It is thus unlikely that rapid evolution will prove to be

Nature's get-out clause for escaping the worst effects of climate change. At least, I wouldn't bet on it. Certainly, I wouldn't bet the diversity of life on Earth on it.

CHAPTER 8

Complex Communities

The aphid and the ladybug

You might be surprised to have reached this far in a book about climate change and biodiversity without having read much about polar bears, whose charisma and cuddly appearance have been widely used by campaigners to captivate the public's attention about the risks of global warming. Well, rest assured that later in this chapter we will explore the polar bear's predicament in some detail, but for reasons that will become clear, we will begin with the relatively unglamorous Pea Aphid—a small, green, rather unattractive insect.

Pea Aphids were introduced to North America from Europe sometime in the 1800s, most likely as unplanned stowaways among plants being transported for agricultural purposes. They have since flourished in the New World, spreading throughout much of the continent, where they feed on peas, beans, alfalfa, and other legumes. Pea Aphids can attain very high population densities and destroy crops, making them a potentially serious agricultural pest. However, in most parts of the United States, they rarely reach densities high enough to cause serious trouble. This isn't due to a lack of crops to provide food, nor is it due to unfavorable climate conditions; rather, aphid populations are kept in check by a suite of natural enemies, including a predatory ladybug, the Asian Lady Beetle.

The Asian Lady Beetle was intentionally released into several states in the United States during the twentieth century as a ploy to control populations of aphids and other pests. Given its economic importance, this particular beetle has been the subject of a great deal of research to understand its precise predatory instincts. Of particular interest to us here is that these beetles are usually absent from fields when aphids are present in low abundance, but they become much more common as aphid abundance increases. In fact, studies have shown that females of the species usually only lay eggs when stimulated by the presence of large numbers of aphids. From an evolutionary standpoint, this behavior makes good sense— the Asian Lady Beetle has evolved to take advantage of high aphid densities, but it reduces predation when prey populations are low, which helps to avoid diminishing the food source.

Ladybug and aphid populations are therefore inextricably linked. When aphids are abundant, ladybugs are abundant; and when aphids are scarce, ladybugs are scarce. This, of course, is similar to the ebb and flow dynamic that is common in interactions between predator and prey: as prey become more abundant, predator survival is enhanced because of the increased availability of food; and then, when predator populations become too high, prey becomes overexploited and both predator and prey populations decrease. You can think of the classic case of foxes and rabbits, whereby fox populations increase and decrease in sync with changes in the number of rabbits. The twist with our ladybug and aphid example is that the ladybug behaves in a way that helps self-regulate its population when aphid densities are low, rather than continuing to exploit the dwindling resource.

Let's now throw climate change into the mix. Pea Aphids are especially susceptible to short periods of high temperatures—heat waves. Temperatures must get exceptionally high before aphids die, but a few hours of unusually hot weather can greatly reduce reproductive success, causing aphid populations to plummet. As mentioned in Chapter 1, heat waves are predicted to become more frequent over the coming century, so it is pertinent to consider how aphids might fare under climate change. In particular, we can ask how the ladybug–aphid dynamic might influence the impact of warming.

A study led by Jason Harmon at the University of Wisconsin–Madison has looked into this question in great detail.[1] Harmon ran a series of experiments whereby field-caged populations of Pea Aphids and Asian Lady Beetles were subjected to artificial heat waves, imposed by covering the cages with clear plastic sheeting for 4 hours every few days. The sheeting increased temperatures within the cages by roughly 9°F, which was enough to reduce aphid reproduction but not so much as to outright kill individuals. In order to unravel the role of the predator in affecting aphid populations, Harmon sometimes included ladybugs in the experiments, and other times he did not.

This suite of experiments demonstrated an important trend: the expected negative effect of the heat waves was ameliorated by a decrease in predation. Basically, as aphid population growth dropped due to the heat waves, predation pressure from the Asian Lady Beetle also dropped, meaning that the impact of the heat waves wasn't quite as bad as might have been expected. By contrast, Harmon repeated the experiments with another predator, the Seven-Spotted Ladybug, which also has a taste for aphids, but which doesn't reduce its predation pressure when aphid numbers are low. In this case, aphids didn't get any respite from predation when heat waves occurred, causing aphid populations to crash much lower than in the experiments with the Asian Lady Beetles. And so, we have evidence that release from predation might, at least in some cases, unexpectedly alleviate the impacts of climate change. Nature is again showing it has hidden tricks for absorbing the impact of environmental change.

A central point for us to take away from the aphid–ladybug example is that interactions between species matter: a species' response to climate change will be influenced by the behavior of other species with which it interacts. Therefore, we cannot expect to understand the impacts of climate change on biodiversity by looking at individual species in isolation; rather, we need to consider species in the context of the ecological communities to which they belong. This notion is the central theme of this chapter, but so far we've only looked at the dynamic between two species—a predator and its prey. I will next describe how climate change can lead to surprising and counterintuitive dynamics across a larger community of interacting species.

A California grassland

The coastal region of California experiences a Mediterranean-type climate, characterized by hot summers with long droughts, and winters that are cool and wet. The distribution of rainfall has an especially important influence on natural systems, with the vast majority of annual rains falling during the winter months. A national assessment of the potential consequences of climate change in the United States, published in 2000, concluded that rainfall is likely to increase substantially in California over the coming century.[2] Of particular interest was a prediction by the Canadian Centre for Climate Modeling and Analysis that suggested the increased rainfall will extend beyond the usual winter season and alleviate the long summer drought. If this turns out to be the case, then we can expect, on the face of it, that plants will flourish under the extended wet growing season.

The question of how California's biodiversity might respond to future changes in rainfall caught the attention of Blake Suttle, who in the year 2000 was about to begin five or so years of doing research toward a PhD at the University of California, Berkeley. Suttle is a field ecologist and had no intention of spending his time as a graduate student sitting behind a computer working on mathematical abstractions of the world outside; no, he wanted to test experimentally, in the field, how nature will respond. Fortunately, the University of California runs an exceptional network of 36 natural preserves that are available for research, and Suttle honed in on one of these—the Angelo Coast Range Reserve in Mendocino County—as an ideal locale for his experiments. Beginning in 2001, Suttle spent many months setting up and maintaining 18 large, circular experimental plots, each with a diameter of

10 meters, in the preserve's grasslands. The plots were rigged so that artificial rainfall could be delivered through sprinklers, the kind commonly used for agricultural irrigation or for watering golf courses.[3]

This experimental setup enabled Suttle to be the weather-maker, adding supplementary rainfall from April through June to extend the growing season. Altogether, the spring supplement increased annual "rainfall" on the plots by around 20 percent. As is necessary with any experiment of this kind, Suttle also left some of his plots alone, without any artificial rains, to act as control plots against which the watered plots could be compared. The one remaining task after having established the experimental design was to painstakingly survey and document changes to the grassland communities—including both plants and insects—for the five-year duration of the study.

So what did Suttle's sterling efforts uncover? For the first couple of years, things progressed pretty much as expected, with plant growth increasing substantially and a lot more species able to coexist in the watered plots. (Plant production, measured as the weight of dry plant material per square meter, more than tripled in the first year.) Importantly, the species that benefitted most in the early years of watering were herbaceous plants, especially those that are able to convert nitrogen from the atmosphere into ammonia, a form of nitrogen that can be used for plant growth. These species—the so-called "nitrogen-fixing forbs," including alfalfa, clover, and peas—play a crucial role in ecological communities because they release nitrogen into the soil when they die. In effect, they are nature's fertilizers.

Some other plant species, specifically the grasses, did not respond to the first year of supplementary rains but increased dramatically in the second year. These grasses did not benefit initially because they complete their annual life cycle too early to take advantage of the artificial spring rains (they were already well on their way to producing seeds by the time the "rains" arrived). However, they did undergo a spurt of growth in the second year—not directly because of the spring rains, but because of the highly fertile soil resulting from the proliferation of nitrogen-fixing forbs the year before. So the grasses did not respond directly to the change in precipitation, but were impacted instead by an indirect effect mediated by the forbs. We therefore find, again, that interactions between species can be crucial in determining responses to climate change.

But now let's look beyond the second year of Suttle's experiment—here's where things become counterintuitive. It turned out that the trend for increased growth and species diversity did not last. From year three onwards, the trend reversed and plant diversity plummeted. The decline was such that by the end of the fifth year, the total number of plant species in the watered plots—the "species richness"—had collapsed to nearly half that in the plots without supplementary rainfall. Biodiversity had taken a significant hit, with the relatively few species that remained being predominantly grasses.

So why would plants suffer when given additional spring rainfall? The explanation for this reversal of fortunes lies in the complex dynamics within the grassland community. We can summarize the full sequence of events as follows: spring watering caused nitrogen-fixing forbs to flourish ... which led to improved soil fertility ... which led to increased growth of grasses ... which caused the accumulation of grass litter (dead plant material covering the ground) ... which suppressed the germination and growth of forbs ... which ultimately led to a decline in plant species richness. Thus, grasses benefited so much from the increased rainfall that they overwhelmed the positive effects of the increased water availability for most other plant species.

If we now shift our focus to the invertebrates inhabiting Suttle's experimental plots, we find a similar pattern of boom followed by bust. For the first couple of years, the plants flourishing in watered plots supported a greater diversity and abundance of invertebrates, including a rich array of aphids, beetles, ants, locusts, crickets, millipedes, and spiders. But as forbs were eliminated in subsequent years, food availability and habitat quality for invertebrates diminished, causing populations to head into steep decline. The problem, it seems, stemmed from the lack of nutritious forbs as food in the summer, and from the lesser structural complexity of habitat offered by a monoculture of grasses when compared with a mixed forb-grass assemblage. As a consequence, the abundance of many invertebrate species was decreased by more than half during the course of the experiment.

All in all, by the end of the five-year experiment, Suttle observed a reduction in invertebrate richness of around 20 percent. This loss of biodiversity did not result from climatic conditions that were inherently detrimental; rather, both animals and plants initially benefitted from the extended growing season, as expected. But as the years went by, species' individualistic responses to climate were overshadowed by the effects of complex interactions within the ecological community. These experiments demonstrate that unforeseen species interactions may be far from inconsequential in controlling responses to climate change; in fact, interactions have the potential to completely *reverse* the response we'd expect.

In moving toward a more complete understanding of how

ecological communities respond to change, we come, at last, to the case of the polar bear.

Compensation for polar bears

Polar bears are one of nature's most charming and alluring predators. In many ways, they have become the poster child of the climate change issue. How many images of a stranded polar bear clinging to the last remnant of a melting ice floe have you seen? Perhaps my favorite example was in January 2009, when, as part of a publicity campaign for a television series, a 16-foot-high sculpture of a polar bear and her cub marooned on a small ice floe was floated down the Thames in London. Here was a stark, media-savvy illustration of the problems faced by wild species.

One of the main reasons to worry about polar bears is the loss of their hunting grounds as Arctic sea ice shrinks year upon year. Polar bears primarily hunt Ringed Seals, which they expertly capture at breathing openings in sea ice or by raiding the seals' subnivean lairs (enclosed dens under the snow layer where they reproduce). It is reasonable to expect that bears will find it more difficult to seize upon their prey when sea ice is scarce—a polar bear certainly has less chance of bagging a seal in open waters. And the problem is likely to be especially severe because early break-up of sea ice at the end of winter coincides with the time of year when a polar bear's energy intake should be at its highest due to the availability of mature seal pups.

Studies of polar bears in western Hudson Bay, Canada,

where break-up of the annual ice is now occurring roughly 2½ weeks earlier than it did in the 1970s, provide evidence that these worries are becoming a reality.[4] Researchers from the Canadian Wildlife Service have carefully monitored polar bears in this region since the 1980s, taking body measurements from individuals tranquilized by shots fired from a helicopter, and following the bear's movements by fitting radio collars that can be tracked via satellite. The research has revealed a decline in the weight, population size, and reproductive success of polar bears in recent years. And as the date of sea ice break-up has become progressively earlier, bears have been forced to leave their hunting grounds and arrive on shore sooner than before; in effect, this is an impact on phenology—an advancing of spring. Usually, the bears will have gorged on seal pups in the weeks before coming ashore, so will be replete and able to get by on berries and vegetation over the summer months. But now, with climate change, it appears that the condition of the polar bears is much poorer.

It is important to note that there is considerable debate among experts as to the relative roles of climate change versus other factors in causing the polar bear declines. For instance, some researchers have argued that increased human–bear contact brought about by a growing tourist industry has had a negative affect on polar bears, perhaps explaining the declining health of the population better than climate change.[5] As we are finding in multiple instances, the most likely explanation is that it is a combination of factors that is affecting polar bear populations.

Let's now focus back on our main interest in this chapter —the role of interactions between species. I want to concentrate on an element of the polar bear story that illustrates a key concept for understanding how ecological communities respond to change—the concept of "compensation." In ecological terms, compensation refers to the possibility that as one link in a food chain is broken, another is created that compensates the loss. It is a bailout from nature.

So, does the lost opportunity for hunting Ringed Seals mean that polar bears will go hungry? Not necessarily. With a bit of luck, the bears will find an alternative food source to feed on.

In Hudson Bay, one alternative food source may be birds' eggs—specifically, the highly nutritious eggs of Lesser Snow Geese. New York researchers Robert Rockwell and Linda Gormezano have shown that it is increasingly likely that over the coming years, polar bears will arrive onshore when members of a large nesting colony of Lesser Snow Geese, located on the Cape Churchill Peninsula, are still incubating eggs.[6] Based on observations dating back to the 1960s, Rockwell and Gormezano found that snow geese are hatching about 1.5 days earlier per decade, whereas bears are forced ashore by ice break-up roughly 7 days earlier per decade. This difference in the rates of phenological change means that under a warming climate, polar bears will increasingly have the opportunity to prey on goose eggs. From the bear's perspective, it is falling out of sync with Ringed Seals, but into sync with Lesser Snow Geese.

And we also know that polar bears will readily take advantage of any opportunity to feast on goose eggs. Bears have been observed consuming eggs in western Hudson Bay and also in Svalbard, a small archipelago that lies about midway between Norway and the North Pole. They often consume the eggs by cracking them open with their noses and then licking out the contents, although sometimes adults will simply eat the entire egg—shell and all.[7]

Given that these eggs are high in protein and fat, it is reasonable to think that this alternative source of food will provide adequate compensation for the deficit in energy reserves brought about by lost seal-hunting opportunities. Indeed, Rockwell and Gormezano compared the energy content of goose eggs and seal pups and concluded that a bear would need to consume the eggs from some 40 nests in order to compensate for one lost day of seal hunting. This kind of rate of egg consumption has previously been observed among polar bears, and with more than 8,000 goose nests on the Cape Churchill Peninsula, there is a plentiful—although not infinite—supply.

Compensation in this way makes communities more resistant to environmental stresses such as climate change, providing somewhat of a safety net for those species that are impacted. We find that food chains are not inflexible: when one link changes (e.g., polar bears–Ringed Seals) it is perfectly feasible that another alternative link will come into play (e.g., polar bears–Lesser Snow Geese). But this is not to suggest that the safety net is infallible, or particularly strong—in fact, as more and more species become extinct from a community, the potential for compensation will be depleted, simply because there are fewer species available to take over a role that is left vacant.[8] In effect, nature can only provide compensation so long as it has resources left in the bank.

And there also remains the problem of indirect impacts on species, like we saw in Blake Suttle's experiments. Here, redistributing links in an Arctic food chain might trigger detrimental impacts on other species in the community. For instance, might polar bears gorge themselves on eggs to the extent that they harm goose populations? Rather like a child let loose in a candy shop, polar bears are known to keep eating and eating when given the opportunity to do so. But unlike a human child, polar bears have remarkably gluttonous capacities, being able to hold up to 20 percent of their body mass in their stomachs. This makes for some extraordinary egg-eating potential, raising the likelihood that Lesser Snow Geese populations could be severely affected by the new predation pressure. In fact, declines in reproductive success among goose populations on Svalbard have already been associated with increased predation by polar bears.[9] And if goose populations are impacted, then we can expect further impacts to reverberate throughout the ecological community.

We are building a picture of ecological communities as highly complex and changeable entities. It is quite possible that some species may be compensated for losses incurred by climate change, making impacts less severe than we might otherwise expect. Yet the potential for indirect impacts within the community means that anticipating future responses is exceptionally tricky. Perhaps, then, we cannot really expect to accurately predict how an ecological community will respond to climate change. However, in the rest of this chapter, we will hone in on a number of general concepts—such as keystone species, tipping points, and phase shifts—that can guide our thinking as to how best to conserve biodiversity under climate change.

Playing Jenga

Ecological communities are often described using the metaphor of a stone arch, in which the stones represent species and the loading forces between the stones represent interactions among species. In such a structure, all stones (species) are dependent on one another for maintaining stability. But all parts are not equal; in particular, the "keystone" is the central piece that has the dominant role in regulating the structure's overall stability. Remove the keystone and the entire structure will come crashing down.

Harvard biologist E. O. Wilson has written about species as being "little players" or "big players" in ecological communities, with the biggest players of all being the keystone species.[10] The most often cited examples of keystone species are large carnivores, which may only be present in low numbers, but which have far-reaching influence through their impacts on a wide variety of prey species and, by extension, on the competitors of their prey. Polar bears are one such example, playing as they do a central role in Arctic communities. The bears not only directly reduce populations of Ringed Seals and other prey through hunting, they also indirectly benefit populations of fish that are preyed on by seals, as well as scavengers such as Arctic foxes, ravens, and gulls, which feed on the carcasses left behind by the bears. So extinction of polar bears from the Arctic system, or a dramatic shift in their diet (to goose eggs, for instance), can be expected to have far-reaching consequences.

This traditional view of ecological communities as comprising more- and less-important species is useful. But ecologists are beginning to reevaluate the keystone metaphor in light of new understanding about how environmental change causes communities to transform over time. Dutch researcher Peter de Ruiter and his collaborators have proposed that the image of a keystone in an arch might be better replaced with the metaphor of the structures built during the popular children's game Jenga.[11] In a game of Jenga, players successively remove pieces from a tower made from a stack of rectangular wooden blocks. As the game progresses, players take turns to remove one block at a time, placing each piece that is removed back on the top of the pile. The Jenga tower constantly changes, and its stability at any point in time is dependent on its current configuration. In this way, any block in the tower can act as a keystone, supporting the weight of the blocks stacked above. The hapless loser, of course, is the player who removes the wrong piece at the wrong time, causing the tower to collapse.

Ecological communities are in many ways similar to a Jenga tower. Unlike a static stone arch, real communities change over time—they are dynamic, with species entering and exiting the community, and links between species coming and going. We've seen that climate change has the potential to rearrange species, assembling new communities as plants and animals shift their ranges and adjust their phenology. The consequences of this reshuffling will be alterations to existing interactions between species as well as the creation of novel sets of interactions. So we should now view ecological communities as dynamic, Jenga-like structures, whereby a species that is a keystone at one point in time may have relatively little influence on the community at another point in time. And vice-versa, a species that is currently a little player may become a big player when circumstances change.

This Jenga-like view of ecological communities tells us something important about what is needed in order to conserve healthy ecosystems into the future. We must strive for "whole ecosystem" approaches to biodiversity conservation, because species that currently appear to be insignificant may play an important role after other species become locally (or globally) extinct.[12] And so, in order to maintain functioning systems, including the natural resilience to change that is afforded by mechanisms such as compensation and release from predation, we need to conserve as many species as possible—not only the big, charismatic species, but also the smaller, seemingly less significant ones.

The popular image of conservation as being all about the polar bears, pandas, and tigers of the world is in some ways misleading. Focusing conservation efforts on charismatic species implies that the risks to society are limited to losing a few attractive creatures. But this misses a key point: functioning ecosystems are necessary for providing essential services to human society (as discussed in Chapter 1), so we must aspire to conserve whole systems rather than a few chosen species. This is one reason why you've had to wait until Chapter 8 to read about the polar bear, and why I started the chapter with the unglamorous Pea Aphid—our story is about the full diversity of life on Earth, including the beautiful and the ugly, the cute-and-cuddly and the not-so-cute-and-cuddly.

That being said, charismatic animals such as polar bears can play an important role as "flagship" (or "umbrella") species in driving conservation action. Typically, protecting the habitat of such species indirectly conserves the entire ecosystem with which they are associated. There is good reason why conservation organizations tend to focus on certain species, but our main concern should be to maintain whole ecosystems. Which brings us to a pivotal question: will entire ecosystems collapse as a result of climate change?

Phase shifts

Here we can extend the Jenga metaphor to include the end point of the game: collapse of the structure. Like a Jenga tower, ecological systems can be subjected to gradual changes that eventually result in sudden, unexpected collapse. Studies of many types of systems—lakes, coral reefs, oceans, forests, deserts—have demonstrated that nature may show little response to gradual changes (for instance, gradual input of nutrients or toxic chemicals, or reduction of groundwater) for a certain time, but may then respond drastically when conditions reach a critical level.[13] This is the "tipping point" that Malcolm Gladwell has made famous.[14]

Climate change, of course, is a relatively gradual change. The worry, then, is that ecological systems will reach a tipping point, at which stage they cease functioning. As climate change causes more and more extinctions, the potential for communities to exhibit compensation is gradually depleted. And as interactions between species increasingly fall out of sync, the potential that a chain reaction will lead to effects across the entire system becomes greater. Eventually, the system will break.

So, much like a collapsed Jenga tower, we may be left with a collapsed ecosystem. However, now the Jenga metaphor begins to lose its usefulness because an important property of real ecosystems is that the old community is likely to be replaced by a new one. Jenga towers don't spontaneously rebuild from new pieces, but in nature, new species will eventually move in and a different way of life will emerge. We are talking now about the idea of a *phase shift*, meaning a transformation from one type of ecosystem to another.

Professor James Brown of the University of New

Mexico has documented how an arid ecosystem in south-eastern Arizona has undergone a phase shift in response to changing climate.[15] Working alongside a succession of collaborators and students, Brown has monitored a 0.2 square kilometer (20 hectare) study plot in the Chihuahuan Desert since the mid-1970s. The landscape in the area is breathtaking, if inhospitable, with the plot itself being roughly 7 kilometers from the dusty hamlet of Portal, whose single store (speaking from experience) provides a limited, yet crucial, selection of beer for visiting ecologists.

When Brown's study was initiated, the study plot was a typical desert shrubland, consisting predominantly of scattered woody shrubs, including acacias, with large barren areas between. But a sequence of aerial photos of the plot (taken in 1958 by the U.S. Forest Service, and then in 1979 and 1995 by a private contractor) documented a dramatic change in the cover of large shrubs through the 1980s. Within little more than a decade, the density of shrubs had increased by three-fold. The most likely explanation for this explosion of shrubs is a shift in the regional climate—specifically, there has been an increase in winter rainfall. It is not surprising that increased rainfall has caused shrubs to flourish, but the most interesting part of Brown's study is how the local animal community reorganized under the new conditions.

Ground surveys of the study plot revealed differing responses across the animal community: a number of species that were previously abundant went locally extinct, at least one new species colonized, and some previously rare species increased their population sizes. Generally, the species that declined were characteristic of arid desert habitats—for instance, the Banner-Tailed Kangaroo Rat, the Silky Pocket Mouse, and two species of harvester ants. Tellingly, all four of these species are *larder hoarders*, meaning that they build up stores of seeds in the summer for eating in the winter, just like squirrels do. It is likely that wetting of these stores due to increased winter rainfall will have damaged the seeds and contributed to the species' declines. By contrast, at least one of the species that increased in abundance—the desert pocket mouse—is less reliant on hoarding seeds, so winter rains will not have been such a problem.

So some species seem to have been affected directly by the changing climate. But it is likely that other species were affected indirectly. For example, the near extinction of the Banner-Tailed Kangaroo Rat, which is regarded as a keystone species in these ecosystems, may well account for substantial regional declines of the Burrowing Owl and the Mojave Rattlesnake, whose populations decreased to as little as one quarter of their original numbers. Both the owl and the rattle-snake make homes in burrows excavated by Banner-Tailed Kangaroo Rats, and the rattlesnake also feeds on these rodents, so loss of the rat most likely contributed to their declines.

The ecological community that Brown first documented in the 1970s, then, had a very different makeup just two decades later—it had experienced a phase shift. However, an important closing to the story is that despite the dramatic changes, the overall diversity of life remained virtually unchanged. As some species declined or went locally extinct, others increased in abundance or colonized from elsewhere, meaning that overall there were roughly as many species present in the community after the phase shift as there were before. In effect, a new Jenga tower emerged where the old one once stood.

All in all, the examples described in this chapter dem-onstrate the complexity and unpredictability of ecological communities. The predictions of future extinction risk that we encountered in Chapter 6 were made using models that cannot adequately account for the multifaceted interactions between species that are crucial for determining responses to climate change. Our confidence in current predictions of future impacts must therefore be reduced.

Yet through this murky outlook, we have been able to identify some general properties of communities—illustrated by the Jenga metaphor—that can inform strategies for effective conservation under climate change. In particular, I have argued the need for "whole ecosystem" conservation. We will explore potential conservation strategies in Chapter 10, but first I want to address the accusation that the problem of climate change is being oversold to the general public. Are environmentalists and scientists sensationalizing the issue?

CHAPTER 9

Crying Wolf?

One million species

Science holds a fairly privileged place in society. Because of its rigorous methods, perceived impartiality, and careful review by outside experts, scientific research is widely given credence above that of pronouncements by political parties or lobby groups. The stereotyped scientist—complete with lab coat, sandals, and socks—is not to be argued with. Respect for scientists is certainly not universal, but it is fair to say that they are generally trusted by the public. However, this trust must be earned and can easily be broken.

Recent controversies in the climate change arena have caused some erosion of the public's trust in science. In particular, the scandal over leaked emails from leading climate scientists that seemed to imply a cover-up—an episode dubbed "climategate"—and controversy over the inclusion of flawed data in the IPCC's Fourth Assessment have raised difficult questions as to the reliability of scientific data. Although independent reviews have cleared the scientists of serious wrongdoing, and have confirmed that there are no errors that undermine the IPCC's main conclusions, public confidence in scientists has undoubtedly slipped.[1]

Against this background, it is pertinent for us to explore how the issue of biodiversity and climate change has been viewed by the broader public. Given the complexity of some of the evidence we've looked at in this book, and the uncertainties that remain, it is tricky for scientists to communicate their research accurately in bite-size chunks that will find their way into newspapers and into television and radio news reports. With reference to the fable of the shepherd boy who cried for help when it was not warranted, some commentators have questioned whether conservation scientists have exaggerated the risks and are "crying wolf" over climate change.

In Chapter 6, I described a number of studies that predict dire consequences for biodiversity under climate change. Foremost among those studies in making it out of the academic literature and into the popular media was the 2004 study lead by Chris Thomas (see page 112).[2] You will recall that Thomas and his colleagues predicted that a potentially substantial number of species in their sample of 1,103 species would be "committed to extinction" based on climate change scenarios for the mid-twenty-first century. The actual proportion of species predicted to go extinct varied greatly—for example, from 11 to 58 percent, depending on the climate change and dispersal scenarios used—but a ballpark, middle-of-the-road summary is that roughly a quarter of the assessed species might be condemned under warming.

Thomas and his team did an excellent job of getting their

research into the media with a hectic schedule of interviews and a carefully worded press release that skillfully packaged the science in a way that would resonate with the public. Here are a few headlines illustrating how the study was reported in the popular press:

> *"Cost of global warming: 1 million species"*
> —*USA Today*
> *"Dire warming warning for Earth's species: 25% could vanish by 2050 as planet heats up"*
> —*San Francisco Chronicle*
> *"Under threat: An unnatural disaster: Global warming to kill off 1m species; Scientists shocked by results of research; 1 in 10 animals and plants extinct by 2050"*
> —*The Guardian (UK)*

Not all newspapers led with such dramatic headlines. For instance, the *New York Times* was rather more measured in leading with: "Scientists predict widespread extinction by global warming." But of thirty-nine newspaper articles that I could get hold of from around the world, more than half included the claim that a million or more species are under threat. And a notable proportion of the articles—about one in four—stated that the extinctions will occur by 2050, even though the original scientific study was careful to clarify that not all predicted extinctions will have occurred by that date (recall the phrase "committed to extinction").[3]

These same embellishments spread elsewhere in the public domain. Of fifty-five Internet sites covering the story, nearly half included the "million species" prediction. Some politicians likewise went with the flow: European Union Commissioner Margot Wallström, at the time responsible for

EU environmental policy, commented on "the recently published study that suggests global warming could wipe out a
third of the planet's species by 2050." The conservation lobby
was also quick to get in on the act, with many of the best-
known environmental organizations—including Greenpeace,
World Wildlife Fund, and Conservation International—
running stories on their websites and producing press releases
and briefing documents. The UK arm of the WWF went so far
as to use the story directly in its fundraising efforts by mail-
ing its members a request for "a special emergency donation"
because of the "shocking news" that "by 2050 global warming
could wipe out one million species of animals and plants."[4]

Clearly, the public face of the study was at least one step
removed from the underlying science. Of course, it is in the
interests of numerous people to liven up a story: media companies sell newspapers and attract increased viewing figures,
environmental organizations generate donations, politicians
align with an issue that may be popular among voters, and
scientists get recognition and funding for their work.[5]

From the scientists' perspective, it is sometimes unavoid-
able that errors creep into media reports. For instance,
despite the best efforts of Thomas and his team to clarify the
time frame for their predictions, many media reports claimed
that the extinctions will occur *by* 2050, thus making the story
a little more punchy and alarming. This error first surfaced
in stories distributed by major news agencies, including
Reuters, whose report dramatized that the extinctions
would occur "by 2050 in one of the biggest mass extinctions
since the dinosaurs."

While some inaccuracies do inevitably find their way into
media reports, scientists themselves must be very careful when
introducing embellishments that are beyond the scope of their

research. It turns out that the "one million species" estimate originated in the press release distributed by the research team. This estimate can be justified on the grounds that it is likely there are at least 4 million species on the planet (which, as you will recall from the opening paragraphs of this book, is toward the lower, conservative end of recent estimates). If we assume the extinction of roughly one quarter of all species—a midrange estimate from the original study of 1,103 species—then we arrive at the target number. This was a clever enhancement of the original study and it proved effective in creating a media splash. Yet the embellishment was beyond the scope of what had been published in the peer-reviewed scientific literature, and it did not pass without criticism. In particular, the "one million species" claim has drawn criticism for exposing conservationists to accusations of crying wolf.[6]

We tend to expect that the media will sensationalize an issue, and that politicians will speak from the perspective of their party line. But the ideal of science is that it remains objective and independent. Embellishing research with claims that can be criticized as alarmist and exaggerated may not only undermine the public's trust in scientists, but could actually prove counterproductive to conservation efforts. There is a risk that claims perceived as being overstated will be met with public cynicism and disbelief—*this is just those tree-hugging greenies spreading scare stories*—that might, in turn, lead to apathy and inaction with regard to tackling the issue.

An informative illustration of this derives from the 2004 disaster movie *The Day After Tomorrow*. The film portrays the calamity that ensues when global warming shuts down the existing circulation patterns in the North Atlantic, leading, paradoxically, to the extremely rapid onset of an ice age—a scenario that has some scientific basis, but is very highly unlikely and is inevitably portrayed in a sensationalized manner. We shouldn't read too much into the film itself because, unlike *An Inconvenient Truth* and *The Age of Stupid*, it was not intended to be a serious comment on the climate change issue—it is simply a Hollywood blockbuster complete with action hero and heartwarming love story. But it is interesting to consider the effect that the movie had on public perceptions about climate change. In Germany, 1,300 moviegoers were asked both before and after seeing the movie whether they believed climate change is real. Interestingly, although the majority of interviewees did think climate change is real, the level of conviction dropped significantly after they had seen the movie.[7] This supports the notion that sensational, exaggerated stories may be counterproductive in efforts to sway public opinion.

Yet breaking a story about biodiversity conservation into the mainstream media is a formidable task, especially given the stiff competition for space from economic woes, terrorism, war, and celebrity antics. Without the "one million species" estimate, it is unlikely that the Thomas study would have been so widely reported, meaning that millions of people around the world would have missed out on hearing that climate change poses a threat to biodiversity. At the end of the day, it is important to ask whether the general picture conveyed by a media report is useful when separated from the complexities and caveats of the original scientific article. As Chris Thomas has pointed out, all that people are likely to have remembered from the media coverage of his study is that climate change could cause a lot of species to go extinct. I agree that this would be a fair representation of the current state of knowledge.[8]

However, as the public debate on climate change moves forward in light of the recent controversies, I believe that more measured and nuanced messages are needed to ensure that public trust in science is maintained. Further, I expect that the scare tactic of invoking catastrophe is likely to prove less effective in instigating behavioral change than are more cautious, easily defendable messages.[9] Of course, measured, caveat-laden conclusions are difficult to get across in a newspaper headline or sound bite. Indeed, as *New York Times* reporter Andrew Revkin has noted, the science of climate change is a bad fit for conventional media—you don't get extra space in a newspaper or time on a broadcast just because the science is so complex.[10] But this should not be used as a reason to avoid gray areas and reduce the debate to black and white viewpoints. As Revkin's reporting ably demonstrates, the ongoing challenge is to communicate the state of knowledge concisely and accurately, avoiding exaggeration and hyperbole.

Problems with scientific advocacy

Most scientists like to be able to show that their research is relevant to societal issues and is important for informing policy makers. A scientist may, for instance, speak out in favor of particular policy decisions, such as international treaties to reduce greenhouse gas emissions. In doing so, researchers risk moving beyond the role of "pure scientists"—they are not only producing new knowledge and understanding, but are also using this knowledge to advocate for particular policy actions. Advocacy is a tricky role for any scientist to take on because it brings into question objectivity and independence. Later in this chapter, I will discuss the necessity for scientists to speak out, but first let's take a closer look at two key ways that scientific advocacy in the climate change debate risks undermining public trust. The first concerns personal values, and the second has to do with gloomy overstatements.

The issue of values can be illustrated with reference to research in which I was involved during the late 1990s. I was working alongside colleagues at the University of Oxford on a project that utilized bioclimate envelope models (very similar to those used by Chris Thomas and his team) to estimate impacts of climate change on plants and animals in Britain and Ireland. As we've now come to expect, our models predicted that some species would spread northward and expand their distributions, while other species, particularly those restricted to highlands in northern regions, would contract their distributions. Overall, we concluded that climate change could substantially alter the biodiversity of the British Isles, with some species being "winners" (for instance, White Beak-Sedge and the Azure Damselfly) while other species would be "losers" (for instance, Alpine Azalea and the Mountain Ringlet Butterfly).[11] Of interest here is that the project garnered a modest amount of media attention, including the following contrasting coverage in the UK press:

> *"Our wildlife in peril: Some of Britain's rarest birds and butterflies will be wiped out within 50 years because of global warming"* —***Daily Express***[12]

> *"A warm welcome: Climate change isn't all bad news for Britain's flora and fauna"* —***The Daily Telegraph***[13]

How can such opposing interpretations be based on exactly

the same piece of research? The answer, of course, lies in personal values and preferences. You might be optimistic about the prediction that some species will be winners from climate change, or you might be pessimistic about the potential for other species to lose out. For instance, nature lovers who enjoy spotting rare birds and butterflies in the Scottish Highlands were likely to be gloomier about our predictions than wine-makers in southern England whose vines would likely flourish under warmer climes. Science cannot tell us that one view-point is right and the other is wrong—it can tell us what is likely to happen, but it cannot tell us whether something is good or bad.

The important point is that interpreting the implications of scientific research requires imposing personal values. By engaging in advocacy, scientists are by necessity imposing their own values. The problem, then, is that scientific findings may be regarded as simply an extension of a scientist's values, lacking in the impartiality and objectivity that is expected of science. If that becomes the case, then public trust will be undermined and scientific information will play a lesser role in influencing public opinion and political debate.[14]

Moving now to the second way that scientific advocacy risks undermining public trust: I will illustrate this point with reference to Professor John Holdren who, as director of the White House Office of Science and Technology Policy, is one of President Barak Obama's top science advisors. Holdren has made some gloomy predictions in the past. In the early 1970s, he warned that some form of ecological catastrophe would almost certainly overtake us before the end of the century, and in the mid-1980s, he boldly predicted that global warming could cause the deaths of one billion people by 2020.

During Holdren's nomination hearing before the U.S.

Senate Committee on Commerce, Science, and Transportation in early 2009, these predictions came back to haunt him when Senator David Vitter pointedly questioned whether they were responsible.[15] Holdren was put on the defensive when quizzed about his "one billion deaths" prediction, admitting that—just one decade short of his 2020 deadline—this was unlikely to happen (although he argued that it was still a possibility, specifically if the global climate system crosses a tipping point that has a catastrophic impact on agriculture). Holdren defended his earlier statements on the grounds that they were not so much "predictions" as "descriptions of possibilities which we would like to avert." He argued that his reason for looking at the more extreme possibilities of what could go wrong was to motivate society to change direction. In effect, then, he was using science to advocate in favor of particular policies.

Although Holdren's nomination was confirmed by a unanimous vote in the U.S. Senate, Vitter's line of questioning targeted a weak spot in the scientist's armor. Pronouncements of impending catastrophe will only maintain credibility if the catastrophe actually occurs, or if a particular policy or action is seen to avert it.[16] You can only use gloomy predictions to motivate change in behavior so many times before people stop listening, as happened to the fabled shepherd boy who cried wolf.

Returning to the example of polar bears (see page 147), if the public is bombarded with the idea that polar bears are about to go extinct due to climate change, and yet they continue to survive by switching their food source, then confidence in claims about the risks of climate change will be undermined.

So, there are significant pitfalls with scientific advocacy, but does this mean that scientists should never speak out beyond the scope of their scientific research? Should scientists retreat to their offices and laboratories, refuse to pick up the phone to journalists, and disengage from any political debate? As I will explain next, I think the answer is a resounding no, but it is essential that the potential pitfalls are carefully avoided.

The need to speak out

Modern society faces a host of complex, controversial issues that rely heavily on scientific input. Examples include not only climate change but also genetic engineering and human embryonic stem cell research. In such cases, scientific information is so central to policy debates that scientists can hardly escape being involved. If scientists ever had the luxury of being able to cut themselves off from societal debate—to lock themselves in their ivory towers—the demands of society are now such that this should no longer be an option.

Moreover, many scientists feel a responsibility to engage with policy debates and the media. Scientists have the privilege of spending their days researching topics such as climate change and they are therefore specially placed to offer informed advice on those subjects that they have studied. And since the taxpayer usually foots the bill, it would be a disservice not to speak out on issues of societal importance. If we also remember that scientists are citizens, with the same right as everyone else to put forward a viewpoint, then it should not be surprising that advice from scientists often covers not only factual understanding but also an interpretation of what should be done.

One prominent voice calling for scientists to speak out is that of Al Gore. In a speech to scientists at the 2009 annual meeting of the American Association for the Advancement of Science (AAAS), Gore had a strong message: "We need you to be actively involved."[17] His principal point of persuasion was that if researchers do not offer their opinions, then others will. For instance, the American Coalition for Clean Coal Electricity, a consortium of major companies from the coal producing, electricity generating, and mining sectors, among others, is spending huge sums on advertising and lobbying to make their position on climate change known. According to Gore: "When they spend $500 million putting their version of this story in the minds of the American people, it increases the importance of you [scientists] being willing to speak out."[18] This is, I believe, an important and valuable argument, and few scientists would disagree that they should spend more time and energy communicating their knowledge to the general public and refuting bogus information.

But Gore does not only want scientists to become better communicators—he wants them to become actively involved in politics. Again from his 2009 AAAS speech: "If I could, I would motivate you to leave this city after this meeting, and start getting involved in politics." On this point, I think we need to exercise caution. The historian Robert Dallek has evaluated the role that academics have previously played within the White House, including under Presidents Roosevelt (Franklin D.), Nixon, and Kennedy. Dallek has noted numerous instances in which political considerations have clouded academics' advice, leading to his conclusion that "professors should confine themselves to what they know and leave the politics to politicians."[19] We must remember that politics involves trying to achieve a preferred outcome through negotiation and compromise,[20] which is at odds with the scientific method and is not something scientists can claim any particular expertise in.

This is not to doubt the immense potential value of distinguished scientists such as John Holdren taking senior positions within the political arena. But in taking on such roles, scientists must be extremely cautious to avoid the pitfalls of scientific advocacy that I have outlined above. It is essential that scientific advisors remain nonpartisan, avoid exaggeration, and admit the shortcomings of current understanding; in short, it is essential that they stick to the science. Outside their "day job," scientists are at liberty to speak out as citizens, but there is a special onus to be clear when they are espousing personal points of view versus when they are speaking about science.

Roger Pielke Jr., a professor of environmental studies at the University of Colorado at Boulder, usefully describes the role that scientists should play in informing political debate as being that of an "honest broker."[21] Such brokers will explore the pluses and minuses of alternative policy options, but stop short of making a value-laden judgment about the course of action that should be taken. An analogy is with a mortgage broker, who, if trustworthy, will inform buyers of the pros and cons of various loan options, but will leave it up to buyers to select the products they decide best suit their needs. In the context of climate change and biodiversity, an honest broker should communicate the kinds of impacts that have already been observed, point to possible future risks while admitting that the future remains highly uncertain, and assess a range of possible policy responses. But the honest broker must leave it up to society and its policy makers to decide what, if anything, they are prepared to do in response. Fulfilling this role of the honest broker is the underlying goal of this book.

In the following, final chapter, we will look at some of the policy and management options that society must decide whether to act on. But first, I will address a central theme of the book by asking whether the message that climate change is an impending disaster is justified or alarmist.

An impending disaster?

I described in Chapter 1 how the language used to talk about climate change is increasingly that of doom and calamity. We have seen this in the media's response to the Thomas study, with newspapers reporting the story using words such as "disaster" and "dire." Is the natural world really facing a climate change catastrophe or are conservationists crying wolf?

Based on the observed fingerprints that we looked at in Chapters 2 to 5—including upslope and poleward migrations, shifts in phenology, and coral bleaching—we have enough evidence to prove wrong the skeptic who denies that climate change is a threat. There is no serious debate over the fact that climate change is already affecting natural systems. However, in Chapters 6 to 8, we have seen the difficulties that are inherent in predicting how the future will pan out. There remain major uncertainties, which cast a doubt on dire predictions from climate doomsayers.

Overall, it is fair to say that, unless major changes in international policies lead to substantial decreases in greenhouse gas emissions, the most likely outcome is that global warming will lead to extensive and irreversible transformations of ecosystems. But whether this will amount to catastrophe—

meaning the collapse of ecosystems and mass extinction—is unclear. My intention is not to sit on the fence on this central matter, but rather to make the important point that the future is uncertain. The fact that we cannot predict how bad the impacts will be is perhaps the most compelling reason for taking action to reduce the risks.

---------- CHAPTER 10 ----------

Twenty-First-Century
Conservation

An anthropogenic cocktail

There has been a dramatic shift in public concern about climate change in recent years. According to surveys of perceptions among the American public, global warming ranked sixth on a list of environmental problems in 2003, lagging behind water pollution, destruction of ecosystems, toxic waste, overpopulation, and ozone depletion. Yet by 2006 it had risen to be the number one environmental concern.[1] This shift in public attention no doubt reflects, in part, the influence of Al Gore's promotion of the issue through *An Inconvenient Truth*, as well as the publicizing of several high profile scientific findings, such as the extinction rates predicted by Chris Thomas and his team. More recent surveys suggest that concern may have waned a little since 2006, perhaps due to the perception that dire warnings have been exaggerated,[2] but climate change remains the most-talked-about environmental issue of our time.

Although there is a tendency to uncouple different environmental problems and rank the importance of individual threats, in reality it is the confluence of multiple factors that is causing the current biodiversity crisis. I am often asked, Why should we be worried about climate change given that the climate has changed many times in the past and species have survived? One response is that contemporary climate change is happening much more rapidly than in the past,[3] meaning that many species, such as Quiver trees, will be unable to respond fast enough. But the answer that I emphasize is that climate change is not the only hazard facing natural systems today. The main problem is that nature is now under threat from multiple quarters, including the destruction and fragmentation of habitats, invasive species, and overharvesting, which together reduce species' ability to respond naturally to the changes in climate.

I have described a number of instances in the book whereby biodiversity is threatened by climate change acting together with other factors. Examples include the interaction between climate change and an infectious disease (chytridiomycosis; Chapter 2) and the combination of warming waters, overfishing, and ocean acidification (Chapters 3 and 5). In such cases, multiple threats amplify the importance of one another, resulting in an overall risk of extinction that is greater than if each of the threats acted independently.[4]

The risk from combined threats is perhaps most apparent if we consider the collective impact of climate change and habitat fragmentation. Modern landscapes are fragmented by a multitude of human constructs—cities, roads, golf courses, agricultural lands, housing estates—that act as impenetrable barriers to species that need to shift their distributions in response to climate change. So while species may have survived climate change in the past by shifting poleward or upslope, they are unlikely to be able to do so in modern landscapes where habitat fragmentation amplifies the risk to species' survival. British biologist Justin Travis has dubbed this combination of threats "a deadly anthropogenic cocktail."[5]

In many regards, the outlook for biodiversity is bleak—we face an uphill struggle to conserve biodiversity as the climate changes over the coming century. But in this final chapter, let us step back from the dire predictions, take a deep breath, and ask what conservation strategies are most likely to be effective. In looking at management options, it is essential that we do not view climate change as an isolated threat, but rather as one threat among many, one ingredient in the cocktail.

Bigger, better parks

Protected areas, such as national parks and nature preserves, are the primary strategy for biodiversity conservation around the world. It is here that wildlife can find sanctuary on a planet dominated by human activities. Yet climate change presents a fundamental challenge even to protected areas: it does not respect preserve boundaries. At the Monteverde Cloud Forest Preserve, for instance, population declines are occurring even though the preserve is protected from outside threats such as deforestation and hunting.

As global warming increases, conditions within parks may become unsuitable for the species they were designed to protect. Daniel Scott, from the University of Waterloo, has illustrated this problem with reference to Canada's Prince Albert National Park, which has a formal mandate to "protect for all time the ecological integrity of a natural area of Canadian significance representative of the southern boreal plains." Scott teamed up with Jay Malcolm to use global vegetation models to predict the stability of the park's vegetation under future climate change. Their results are somewhat sobering: the models projected the eventual loss of boreal forest, suggesting that the park's mandate is untenable in the long term.[6]

However, there are good reasons to think that protected areas will remain our best bet for conserving biodiversity over the coming century. By reducing nonclimatic threats, parks and preserves are able to maintain ecosystems that have a diversity of species and healthily sized populations. As we've seen, diverse ecosystems are more resilient to climate change due to properties such as compensation, while large populations have high genetic variation and consequently increased capacity for rapid evolutionary adaptation. Reducing nonclimatic threats is therefore essential for helping to buffer the effects of climate change. Ecosystems that are protected from other threats—from the anthropogenic cocktail—will be best placed to meet the climate challenge.

Moreover, the value of parks and preserves could be further enhanced if we are smart about which new tracts of land are selected for protection. High priority should be given to conserving areas that exhibit as much environmental variation as possible, giving species the opportunity to shift their distributions within the park as conditions change. For example, wherever possible we should protect habitats that cover a range of elevations, to enable species to shift their distributions upslope as the climate warms. This is especially relevant in Madagascar, where the government has announced a major initiative to select new areas for protection.[7] Chris Raxworthy's work in the northern highlands gives credence to the proposal that intact habitat on major massifs will be a particularly smart choice.[8]

It is also the case that, as conditions in a preserve become unsuitable for one set of species, conditions will become favorable for a different set of species. Ecological communities within protected areas may undergo phase shifts as one set of species leaves and an alternative set of colonizers arrives.[9] This possibility will require a change in the way protected areas are managed. Whereas conservation management within a particular park tends to focus on preserving the ecosystem that is currently there, policies of the future should see nature as dynamic and must accommodate change. For instance, the long-term value of Prince Albert National Park should not be judged on its success in conserving boreal forest, but rather on its success in conserving a healthy, diverse ecosystem—whatever that ecosystem may look like.

The potential for ecosystems in parks to undergo phase shifts raises another important issue: In order for previously absent species to colonize, it will be necessary for them to shift their ranges into the park. This is a tall order given modern fragmented landscapes. An important advantage of creating more protected areas will be that the distance between them decreases; in this way, parks become more connected, with species having a better chance of shifting their distributions by using "stepping stones" of protected habitat. But, inevitably, species will be required to pass through human-dominated lands in order to shift their ranges. Another key challenge for conservation planning is therefore to help species to transit from one protected area to another.

Safe transit

One widely discussed proposal for facilitating species' range shifts is commonly termed *assisted colonization*. The concept is simple: in cases where species are unable to naturally relocate as the climate changes, either because of inherently poor dispersal ability or because artificial barriers such as roads stand in their way, we collect them up, put them on a truck, and artificially move them to a new location. The process is akin to Ian Woodward's transport of navelwort seeds around Britain (see Chapter 7), except the aim here is to establish new populations for conservation purposes, rather than for scientific study.

Pilot studies have already demonstrated the feasibility of this approach, at least for some species. For instance, in 1999 and 2000, British ecologist Steve Willis led a team of researchers who relocated populations of two species of butterfly—the Marbled White and Small Skipper—to sites poleward of their existing ranges. Around 500 individuals of each species were netted at sites in northern England and then transported a few dozen miles north to two previously uninhabited nature preserves where they were released. The newly established populations took hold and were still thriving in 2008, giving credence to the idea that the approach could be rolled out more widely to facilitate climate-induced range shifts.[10]

The potential to utilize assisted colonization appeals as a somewhat last-ditch measure to move species to a new safe haven as the climate changes. However, the strategy has a number of serious drawbacks. First, there are many instances whereby well-intentioned species introductions have gone awry, causing unintended damage to the invaded community.[11] Ecological communities can be highly complex and unpredictable (think back to Chapter 8), so we should be very cautious about our ability to engineer healthy ecosystems by artificially introducing new species.

Assisted colonization is thus a somewhat risky strategy, yet it is reasonable to think that careful study of individual cases could identify situations where the potential benefits outweigh the risks. Such situations would involve relocating species over relatively short distances to sites that share broadly similar groups of organisms. (The idea is certainly not to relocate herds of rhino to the southwestern United States, nor to move polar bears from the Arctic to Antarctica!)[12] The chances of successful colonization may also be enhanced by using bioclimate envelope models (see Chapter 6) to identify sites that will become increasingly suitable for the species under future climate change.[13]

Assisted colonization may therefore prove successful in carefully planned instances. But herein lies a second, and more important, drawback of the approach: it is not realistic to expect that assisted colonization could be safely applied to more than a tiny fraction of the species that are threatened by climate change. The resources required to plan and carry out relocations will restrict this strategy to a small number of (most likely charismatic) species.

So what can be done to help more species shift their distributions? We need to start thinking about facilitating natural dispersal of species through lands that are not formally set aside for nature. That is, we need to find ways to make the land outside protected areas more hospitable to nature. For example, corridors of natural habitat running through human-dominated landscapes, including over- and underpasses across roads, can enable some species to move between protected areas.[14]

Agricultural land occupies nearly 40 percent of the planet's land surface, so making these areas more hospitable to nature is especially important (particularly given that global food demand is expected to double over the next 50 years).[15] Many initiatives—known as *agri-environment schemes*—are already in place to encourage farmers to promote biodiversity. For instance, more than $5.5 billion are paid annually through agri-environment schemes run by the U.S. Department of Agriculture and the European Union's Common Agricultural Policy. These funds are paid to farmers for measures such as creating and restoring woodland, grassland, and ponds on their farms, and for providing food for farmland wildlife. Such initiatives show promise for providing benefits to biodiversity, and increasing these schemes is likely to be a valuable strategy in responding to climate change.

Thus, our overall vision for twenty-first-century conservation should be to create landscapes that allow species to move freely and naturally between protected areas. We must begin to view nature as dynamic and changing, rather than as static and amenable to preservation in disconnected fragments of habitat. This vision is no pipe dream; in fact, there are a number of promising initiatives that demonstrate the feasibility of the approach.

The Yellowstone to Yukon—or Y2Y—Conservation Initiative is one such example. The Y2Y region runs roughly 3,200 kilometers from the Mackenzie Mountains of Canada's Yukon and Northwest Territories to the southern end of Yellowstone National Park in Wyoming.[16] This vast tract of land—an area larger than California and Texas combined—incorporates numerous protected areas, including Yellowstone, Banff, Glacier, and Jasper National Parks, as well as towns, fields, and lands owned by diverse private landowners. The Y2Y initiative comprises a network of more than 300 conservation groups that are working together to create a landscape that maintains enough connected habitats to allow species to move between protected areas. By looking strategically at the region as a whole, priority can be given to supporting conservation schemes that create and maintain connectivity between parks and preserves. And because the region is oriented north–south along the Rocky Mountains, it is ideally situated to facilitate poleward range shifts under climate change.

Comparable initiatives are under development around the world, including the Great Eastern Ranges Initiative in Australia, and the Mesoamerican Biological Corridor project.[17] Such schemes should become a mainstay of conservation efforts through the coming century. Yet despite our best efforts, some species will inevitably remain imperiled—we're thinking now, for instance, of frogs caught up in the global amphibian crisis, or species isolated on mountaintops with no prospect of shifting upslope to higher ground. The only hope under some circumstances may be to preserve such species in captivity, outside their natural environment, using an approach known as *ex-situ* conservation.

Ex-situ conservation

There is a long history of maintaining species in zoos, aquariums, botanical gardens, and seed banks, so methods for captive breeding and plant propagation are well known. In many ways, *ex-situ* conservation is analogous to Noah's ark: species facing extinction are stored in captivity, as if on board an ark to escape the flood, until the threats to existence are eliminated and they can be reintroduced into their natural environment.[18]

One such project is the Amphibian Ark, a coordinated effort between the world's zoos and aquariums to support captive breeding of amphibian species. In response to the global amphibian crisis, conservationists have started gathering up frogs that seem destined for extinction and rearing them in captivity. The Amphibian Ark initiative is raising money to support captive breeding programs all around the world, with an overall goal of transferring 500 individuals of 500 species into protective custody within 5 years.[19] This may prove invaluable as a desperate, last-ditch attempt to conserve an entire class of species. Indeed, it may be the only hope for some of the imperiled amphibians of Monteverde.

Seed banks are another example of *ex-situ* conservation. The Millenium Seed Bank at Kew Gardens, UK, stores more than 3 billion seeds that represent some 10 percent of the world's known flora. Provided sufficient funding can be maintained, the target is to up this to 25 percent of all plants by the year 2020.[20] Seed banks are an important source of seeds for replanting damaged lands, and give scientists the opportunity to search for potential medical uses. These vast collections may also prove to be an important resource for identifying plants that are best adapted to thrive on a warmer planet, such as varieties that are especially tolerant of drought. This is likely to be especially valuable for agricultural crops. For instance, the International Rice Research Institute in the Philippines stores more than 100,000 strains of rice in a cryogenic gene bank, providing a resource that may be invaluable for adapting one of the world's most important cereal grains to climate change.[21]

So, *ex-situ* conservation provides some vital opportunities.

However, the strategy also suffers from some important limitations. Cost is one major consideration. Plant seeds are relatively cheap to collect and store, meaning that a significant portion of global plant diversity can be gathered together, as proven by the Millennium Seed Bank, at an average cost of roughly $2,800 per species.[22] However, captive breeding of animals is an extremely expensive business. For instance, the estimated five-year running cost of the Amphibian Ark is $50 million, which means the price tag per species runs at around $100,000.

Ex-situ conservation thus suffers the same limitation as assisted colonization: the number of species that we can expect to save is only a small fraction of those threatened by climate change. In Chapter 8, I argued that the focus of management efforts should be on conserving whole ecosystems rather than a few charismatic species, since it is the natural diversity of ecological systems that can make them resilient to climate change and other threats. In light of this, *ex-situ* conservation and assisted colonization—both strategies that operate on a species-by-species rather than whole-ecosystem basis—should be regarded as last resort measures, rather than as mainstays of efforts to conserve biodiversity over the coming century.

The root of the problem

The topic of *ex-situ* conservation also brings us to another crucial point: So far, I have focused on efforts to adapt conservation strategies to climate change, but it is essential for biodiversity conservation that the root of the problem—climate change itself—is slowed and ultimately halted over the coming century. This is essential for *ex-situ* conservation because success of the strategy is dependent on species one day being reintroduced into the wild—yet it is by no means clear when or where this might be feasible unless changes to the climate system are addressed. As things stand now, species stored *ex-situ* will essentially become living fossils,[23] with little chance of ever being reintroduced to play a functional role in natural ecosystems.

Curtailing the rate and magnitude of climate change will be essential not only for the success of *ex-situ* conservation, but also for each of the other conservation strategies I've described. There is no getting away from the fact that in many cases the magnitude of the problem facing natural systems is likely to overwhelm conservation efforts unless climate change is reined in.

This can be starkly illustrated by considering the magnitude of the problem facing coral reef conservation efforts. There are numerous initiatives that show potential for restoring bleached reefs. For instance, at Sekisei Lagoon in the East China Sea, the Japanese government is investing heavily (to the tune of around $430,000 per year) in efforts to restore the reef using specially designed ceramic disks with grooves that coral larvae can nestle in. Divers are placing cages filled with the disks in surviving parts of the reef and then, once colonies have started to form, moving the disks to bleached areas in order to kick-start recovery. In a more low-tech example, the coastal community of Bolinao, Philippines, has demonstrated the potential to help recovery of the reef simply by breaking off fragments of coral from healthy parts of the reef and wedging the pieces into cracks in bleached sections. During a training workshop in 2008, Bolinao fishermen replanted about 400 square meters of reef, 80 percent of which was growing well 6 months later.[24]

Conservation efforts such as these will have value for aiding the recovery of reefs at a local scale. However, the feasible scale of reef restoration (a few square kilometers per year) is dwarfed by the extent of degradation (many hundreds of square kilometers).[25] Moreover, the reefs are likely to take decades to recover. The magnitude of the problem is therefore such that the future of corals is largely dependent on tackling the warming and acidifying waters that are the root cause of the problem.

There are many good books that discuss ways to mitigate climate change, including through international legislation—in particular, the need for a legally binding and effective follow-up to the Kyoto Protocol, which failed to be achieved at the 2009 UN Climate Change Conference in Copenhagen—and a shift toward low-carbon energy sources such as wind, solar, and/or nuclear power.[26] An in-depth discussion of options for reducing the atmospheric concentration of greenhouse gases is beyond the scope of this book.

However, it is important to emphasize here that nature conservation will not only benefit from curtailing climate change—it is also an essential part of the solution to climate change. Ecosystems have an important influence on the global climate system by pulling huge quantities of carbon out of the atmosphere and storing it in living matter. Destroying ecosystems releases carbon back into the atmosphere, which is itself a significant driver of climate change. This is perhaps most apparent if we consider the world's forests, especially the tropical rainforests of Amazonia, which have been very rapidly destroyed in recent decades, primarily for expansion of cattle and soybean production.[27] In Brazil, deforestation accounts for as much as 70 percent of the country's total greenhouse gas emissions while, worldwide, deforestation accounts for around 20 percent of all emissions.[28] This is a larger share of global emissions than is produced by the entire transport sector.[29]

Increasing the size and quantity of protected areas, especially in tropical rainforests, is therefore not only a key strategy for conserving biodiversity under climate change, but is also a key strategy for tackling the root cause of the problem. It is a win-win situation, but, of course, halting deforestation and vastly increasing the amount of protected area is no small task. Which brings us to our next key questions: How much would it cost? And who pays?

The price of conservation

Existing protected areas cover roughly 8 percent of the Earth's land and 0.5 percent of the Earth's ocean area. The money spent globally on these parks and preserves totals around $8 billion per year. These costs are largely for operating budgets for maintenance, staffing, and control of unsustainable activities such as poaching.[30] Suppose we were to target increasing the amount of protected area to incorporate 15 percent of the land surface and 30 percent of waters. University of Cambridge Professor Andrew Balmford, working with a large team of collaborators, has calculated that the total cost of such a scheme would be in the ballpark of $53 billion per year for the next few decades. Much of this cost would be for compensating landowners when lands are turned into preserves, and for strict law enforcement to ensure that the preserves truly protect nature.[31]

Any cost measured in billions of dollars is difficult to get our heads around, so it is worth putting the price tag associated with expanding protected areas into perspective. For instance, worldwide stimulus plans to boost the floundering global economy totaled nearly $3 *trillion* at the beginning of 2009.[32] So in the grand scheme of things, and considering what is at stake, protected area costs might be considered somewhat of a bargain.

Financing nature conservation is problematic because the economic value of services provided by ecosystems is rarely accounted for appropriately. Recycling of wastes, purification of drinking water, maintenance of soil fertility, storage of carbon, and protection from coastal storm surges are all examples of services that we consider "public goods"—they are available to everyone and do not come with a price tag. Forestry provides an example: a logging company might pay for the right to cut down trees, but no one picks up the tab for the resulting loss of carbon storage. Another example is provided by the 2010 Gulf of Mexico oil spill. Although BP has been required to put $20 billion in a trust fund to pay for clean-up costs and compensation for local businesses, it is likely that the true cost of the loss of ecosystem services, including the carbon storage and protection from hurricanes that are provided by marshes, will run a great deal higher than $20 billion and will never be accounted for.[33] This is a fundamental failure of economics because neither the finiteness nor the fragility of ecological systems is recognized.[34]

Climate change might be a game-changer. More than any other environmental issue before, climate change is demonstrating to people the connections between the natural environment, the economy, public health, and agriculture. Here is an issue that links biodiversity conservation, burning fossil fuels, coastal flooding, forest fires, hurricanes, insurance premiums, disease outbreaks, crop yields, and even the long-term prospects for popular ski resorts. In this regard, climate change offers huge opportunities for putting an economic value on ecosystem services and financing biodiversity conservation.

New financing initiatives to conserve forests are now a serious possibility through international markets in emissions that were generated by the Kyoto Protocol, the international treaty that set targets to reduce greenhouse gas emissions over the period 2008–12. Under the treaty, countries must meet their targets primarily through national measures, but it is also possible for countries to gain credit by way of market-based mechanisms. More wealthy countries can offset some of their emissions by funding projects that reduce emissions in less wealthy countries, and also by buying and selling emissions in what have become known as "carbon markets."

For a host of political and practical reasons, the carbon currently stored in forests was excluded from dealings under the Kyoto Protocol. However, it is likely that this is about to change. Forest preservation could be funded by permitting wealthier nations to gain carbon credits by paying for initiatives that reduce deforestation in other countries, or by allowing the carbon stored in uncut trees to be traded in the carbon market. This is a proposal known formally as REDD: Reducing Emissions from Deforestation and Forest Degradation.

Carbon emissions trading is now a very big and rapidly expanding business. The total value of global carbon markets was around $126 billion in 2008, which is double that of the year before,[35] and by some estimates the market's value will be trillions of dollars within a decade.[36] Allowing carbon stored in forests to become part of this marketplace has the potential to generate significant funds for forest conservation. These funds could be used to ramp up existing initiatives to reduce deforestation rates, including through expansion of protected areas and efforts to tackle illegal logging. Smart investments would also help to modernize local agriculture so that yields can be increased on land that is already cleared, reducing the demand for deforesting more land.[37]

The need to move forward with REDD was one of the few points of agreement at the 2009 Copenhagen conference, so it is probable that a mechanism of this type will be central to any successor to the Kyoto Protocol.[38] If REDD can get the green light, then perhaps similar approaches could be developed to give value to other public goods.[39] Much is being made of possible new technologies for capturing carbon from the atmosphere and storing it—say, by building gigantic pumps that draw in air and filter out the carbon dioxide for storage underground[40]—yet natural ecosystems provide a method for capturing and storing carbon that has been tried and tested over millennia.[41] So why not better recognize the value of ecosystem-based approaches?

It will take a great deal of effort and skill to devise and successfully implement policies such as REDD. However, good ideas for valuing ecosystems and conserving biodiversity are close at hand. It can be done. And it makes economic sense to do so. What is needed is the social and political will to make it happen.

How much risk are you willing to take?

You will recall that in the opening chapter, President Barack Obama was quoted on the need to tackle climate change: "we risk consigning future generations to an irreversible catastrophe." The key word here is *risk*.

There are two components to the concept of risk, namely the *likelihood* (or *chance*) that an event will occur, and the *magnitude* of the consequences of that event. Dealing with risk is an essential part of life. For instance, if the meteorologist predicts a 60 percent chance of rain, you may still choose to light the barbecue because it wouldn't be too calamitous if the meal had to be rushed back indoors. But if the prediction is a 20 percent chance that a hurricane will strike, the potential magnitude of the event may well make you board up and head inland.

A starting point for the book was the IPCC's conclusion that it is *likely* (meaning there is at least a 66 percent chance) that climate change will lead to the extinction of a substantial proportion of species (at least 1 in 5). We have now seen how extraordinarily difficult it is to put numbers on the likelihood and magnitude of future extinctions due to climate change, but we know the possibility of massive species loss is out there. The future is inherently unpredictable, leaving us to deal in terms of probabilities, uncertainties, and varying levels of confidence as we plan for the coming decades. It is not possible to predict precisely how this game of chance will pan out, but we do know that climate change is loading the dice against species' survival.

Through all this uncertainty, society must make decisions about what action we will take to conserve biodiversity and tackle climate change. Science cannot tell us what to do.

Decisions over policies such as how much to invest in protected areas, or whether to introduce a REDD-type scheme, are not scientific decisions—they depend on a trade-off between the cost of acting versus the risk of not acting. Scientists can describe our best understanding of the issue, but they cannot dictate what action should be taken. That depends on personal values and the risks people are prepared to take.

We reach the point where you, the reader, armed with a thorough understanding of the issue, must form a personal opinion as to what action society should take. Your opinion matters. You have the power to influence society by exercising your democratic right to vote and by spending your money on products and services that are in tune with your values. (For instance, are you willing to pay a surplus on your electricity bill for sustainable energy, or perhaps additional tax to fund bigger, better protected areas?) Ultimately, it is public opinion and consumer demand that steer society.

I have tried to describe accurately the current state of knowledge on this issue. Now it is up to you as a citizen to help chart our course for the future.

NOTES

Full reference information for books and articles cited in these Notes can befound in the References, pages 208–222.

Chapter 1

[1] It is tricky to define exactly what a "species" is. The most commonly used definition—known as the "biological species concept"—says that a species consists of a group of organisms that can successfully interbreed with each other, but not with other species. However, this is not the only possible definition, and it does not account for the large number of organisms that do not reproduce sexually. Furthermore, it is very difficult in practice for biologists to find out whether two similar groups of organisms, which may be found in different areas, are potentially capable of interbreeding. Put most simply, a species consists of organisms that are similar and closely related.

[2] Estimates regarding the number of species discovered each year and the total number of species are based on R. M. May (2010), A. E. Magurran (2005), and E. O. Wilson (2001).

[3] D. Raup (1991).

[4] A detailed account of Vietnam's climate and biodiversity is provided by E. J. Sterling *et al.* (2006).

[5] A. Balmford *et al.* (2002).

[6] F. S. Chapin III *et al.* (2000).

[7] Al Gore makes this third point well in the film *An Inconvenient Truth* by likening society's response to that of a frog placed in a pot of water: if the water is already boiling the frog will immediately jump out of the pot, but if the water is initially warm and then slowly brought to the boil, the frog will not react and will be boiled alive (unless rescued, as in the film!).

[8] The IPCC was especially under fire toward the end of 2009 when leaked emails from scientists on the panel suggested that data was being withheld from climate change skeptics and that there were efforts to exclude contradictory findings from influential reports (a scandal widely dubbed "climategate"; see A. Revkin, 2009). The IPCC was also criticized in early 2010 when it was revealed that figures it presented concerning the rate of melting of Himalayan glaciers were incorrect (see E. Rosenthal, 2010). Independent reviews have since cleared the IPCC and its scientists of any serious wrongdoing, and have concluded that the panel's main conclusions remain robust (see M. Russell *et al.*, 2010, and Netherlands Environmental Assessment Agency, 2010). See also Chapter 9.

[9] Intergovernmental Panel on Climate Change Fourth Assessment Report, released in 2007 and consisting of four volumes, which are available at: www.ipcc.ch. The main findings are summarized in IPCC (2007a).

[10] The IPCC's conclusion regarding extinction risk (as set out on pages 13 and 14 of IPCC, 2007a) is further qualified by a note that there is *medium confidence* in this forecast (meaning that the assessed chance of the finding being correct is about 5 out of 10).

[11] Speeches made at the United Nations Climate Change Summit, New York, September 22, 2009. Full text of President Obama's speech is available at: www.whitehouse.gov/the_press_office/Remarks-by-the-President-at-UN-Secretary-General-Ban-Ki-moons-Climate-Change-Summit. Full text of Secretary-General Ban Ki-moon's speech is available at: www.un.org/News/Press/docs//2009/sgsm12470.doc.htm.

[12] Speech made at the Major Economies Forum, London, October 19, 2009. Full text is available at: www.number10.gov.uk/Page21033.

[13] The idea of a climate "fingerprint" has previously been used by C. Parmesan & G. Yohe (2003).

Chapter 2

[1] N. Myers *et al.* (2000). For updated information on biodiversity hotspots, see www.biodiversityhotspots.org.

[2] S. Goodman & J. P. Benstead (2003).

[3] C. J. Raxworthy *et al.* (2008).

[4] IPCC (2007a).

[5] Much of the general information in this section on the Monteverde cloud forest was gleaned from N. M. Nadkarni & N. T. Wheelwright (2000).

[6] J. M. Savage (2000).

[7] J. A. Pounds, pers. comm.

[8] J. A. Pounds *et al.* (1999).

[9] S. M. Stuart *et al.* (2004).

[10] J. A. Pounds *et al.* (2006). The idea behind the climate-linked epidemic hypothesis was introduced in J. A. Pounds & M. L. Crump (1994).

[11] Formally, Harlequin Frogs belong to the genus *Atelopus* and are true toads.

[12] The database used by J. A. Pounds *et al.* (2006) was compiled and first analyzed by E. La Marca *et al.* (2005).

[13] S. Ron *et al.* (2003).

[14] J. Voyles *et al.* (2009).

[15] See, for instance, K. R. Lipps *et al.* (2008), J. A. Pounds & L. A. Coloma (2008), and J. R. Rohr & T. R. Raffel (2010).

[16] J. A. Pounds (2001).

[17] S. M. Whitfield *et al.* (2007).

[18] J. A. Pounds (2001).

[19] A. R. Blaustein & A. Dobson (2006).

Chapter 3

[1] B. Huntley (2005).

[2] S. T. Jackson & C. Weng (1999).

[3] M. B. Davis & R. G. Shaw (2001).

[4] B. Huntley & H. J. B. Birks (1983).

[5] B. Huntley (2005).

[6] S. Levitus *et al.* (2005).

[7] IPCC (2007b).

[8] The information in this section on British coasts is mostly from A. J. Southward *et al.* (1995, 2005).

[9] S. J. Hawkins *et al.* (2003).

[10] A. J. Southward & D. J. Crisp (1956).

[11] Some of the species listed have expanded their ranges eastward, rather than northward, into cooler waters of the eastern English Channel. Data from S. J. Hawkins (2006).

[12] A. L. Perry *et al.* (2005).

[13] A. J. Southward *et al.* (1995).

[14] A. T. Hitch & P. L. Leberg (2007).

[15] C. D. Thomas & J. J. Lennon (1999).

[16] C. D. Thomas *et al.* (2006).

[17] W. Foden *et al.* (2007).

[18] "Auto-amputation": Wendy Foden, pers. comm.

[19] N. Myers *et al.* (2000). For updated information on biodiversity hotspots, see www.biodiversityhotspots.org.

Chapter 4

[1] The comment that record-keepers tend to favor spring events over fall events has been made by C. Parmesan (2006).

[2] The Marsham family record is described and analyzed in T. Sparks & P. Carey (1995).

[3] N. L. Bradley *et al.* (1999).

[4] Examples in this paragraph are drawn from F-M. Chmielewski *et al.* (2004), R. Ahas (1999), A. H. Fitter & R. S. R. Fitter (2002), L. J. Beaumont *et al.* (2006), E. G. Beaubien & H. J. Freeland (2000), S. G. Taylor (2008), and G. V. Jones & R. E. Davis (2000).

[5] X. Chen *et al.* (2005).

[6] C. Both & M. Visser (2001).

[7] M. Visser *et al.* (2006).

[8] C. Both *et al.* (2006).

[9] M. Visser & C. Both (2005).

[10] T. Sparks & P. Carey (1995).

[11] M. Visser & L. Holleman (2001).

[12] M. Edwards & A. J. Richardson (2004).

[13] G. Beaugrand *et al.* (2003), Q. Schiermeier (2004a). To delve still further into the complexities, including the role of distribution shifts among zooplankton, see A. J. Richardson (2008).

Chapter 5

[1] M. D. Spalding *et al.* (2001).

[2] The basic biology of corals is described well in O. Hoegh-Guldberg (1999).

[3] C. R. C. Sheppard (1999).

⁴ For a description of coral bleaching, take a look at J. Reaser *et al.* (2000).

⁵ Further information about the Global Coral Reef Monitoring Network can be found at: www.gcrmn.org.

⁶ The 2002 Great Barrier Reef bleaching event is described in R. Berkelmans *et al.* (2004). The 2005 Caribbean bleaching event is described in C. R. Wilkinson & D. Souter (2008).

⁷ An extensive account of the 1997–1998 coral reef bleaching event is provided in C. R. Wilkinson (2000).

⁸ Coral recovery following the 1997–1998 bleaching event is described in C. R. Wilkinson (2004).

⁹ As mentioned in Chapter 1, information on temperatures in the distant past can be deduced from ice cores. The numbers reported here are taken from O. Hoegh-Guldberg *et al.* (2007).

¹⁰ An introduction to El Niño and La Niña is provided by the U.S. National Academy of Sciences (2000).

¹¹ C. R. Wilkinson *et al.* (1999).

¹² O. Hoegh-Guldberg (1999).

¹³ O. Hoegh-Guldberg *et al.* (2007).

¹⁴ See, for example, S. Catovsky & F. A. Bazzaz (1999).

¹⁵ See, for example, J. M. Hall-Spencer *et al.* (2008), and F. Gazeau *et al.* (2007).

¹⁶ K. E. Carpenter *et al.* (2008).

¹⁷ A. E. Derocher *et al.* (2004).

¹⁸ Y. Yom-Tov & S. Yom-Tov (2004).

¹⁹ Z. G. Davies *et al.* (2006).

²⁰ C. Parmesan & G. Yohe (2003), T. L. Root *et al.* (2003), and C. Rosenzweig *et al.* (2008).

²¹ The number of "species" assessed by Parmesan and Yohe actually includes some categories of species, or "functional groups" (for example, marine copepods).

²² The terms *very high confidence* and *high confidence* here mean that the assessed chance that a finding is correct is 9 out of 10, and 8 out of

10, respectively. Confidence is lower in marine and freshwater systems largely because far fewer records are available from these systems, and also because of the difficulty of distinguishing climate impacts from other stresses (for example, overfishing). IPCC (2007a).

[23] IPCC (2007a).

Chapter 6

[1] This reasoning about species' tolerance based on past climates was put forward by R. Colwell *et al.* (2008).

[2] C. Musil *et al.* (2005).

[3] My description of global vegetation models (GVMs) refers to the type of model used in J. Malcolm *et al.* (2006). My focus on this type of GVM is because subsequent paragraphs describe the results from Malcolm *et al.*'s study. However, it is important to note that newer versions of these models—termed dynamic global vegetation models (DGVMs)—function somewhat differently; for a review of DGVMs, see I. C. Prentice *et al.* (2007).

[4] J. Malcolm *et al.* (2006).

[5] The species–area relationship is described by the formula $S=cA^z$ where S is the number of species, A is the area, and c and z are constants. The analysis by Jay Malcolm and his colleagues, Malcolm *et al.* (2006), also applied a variant of this formula, known as the "endemic–area relationship," which takes into account the number of species expected to be confined to small patches.

[6] For a review of the bioclimate envelope approach, see R. G. Pearson & T. P. Dawson (2003).

[7] For further information on some of the methods that have been used to model bioclimate envelopes, see J. Elith *et al.* (2006).

[8] W. Foden *et al.* (2007).

[9] Mapping of protea distributions has been undertaken by the Protea Atlas Project: http://protea.worldonline.co.za.

[10] G. F. Midgley *et al.* (2002). See also L. Hannah *et al.* (2005).

[11] A. T. Peterson *et al.* (2002). Note that this paper reports bioclimate envelope modeling for over 1,800 species in Mexico. However, here I only discuss results for the 334 species that are endemic to Mexico, because predictions for the other species modeled are biased by political boundaries. (Their bioclimate envelopes were artificially curtailed because species' distributions outside Mexico were not available.)

[12] W. Thuiller *et al.* (2005).

[13] S. E. Williams *et al.* (2003).

[14] C. D. Thomas *et al.* (2004).

[15] These changes are global means.

[16] Chris Thomas and his colleagues tested three different ways of calculating area losses for applying the species–area relationship: 1) change in area summed across all species; 2) proportional loss of area averaged across all species; and 3) change in area for each species individually (i.e., loss of species calculated based on change in area for each species and then averaged across all species). The results presented in the main text are averages across the three approaches.

[17] The information regarding reporting in the media and by politicians is from R. J. Ladle *et al.* (2004, 2005), and L. Hannah & B. Phillips (2004).

Chapter 7

[1] E. Weber & B. Schmid (1998).

[2] E. Weber (2001).

[3] F. I. Woodward (1990).

[4] My use of the term "resurrection ecology" follows M. B. Davis *et al.* (2005).

[5] J. C. Lerman & E. M. Cigliano (1971).

[6] S. Sallon *et al.* (2008).

[7] M. C. Vavrek *et al.* (1991).

[8] IPCC (2007a).

[9] Further examples demonstrating that rapid adaptation can occur are reviewed in M. B. Davis *et al.* (2005).

[10] The Pitcher Plant Mosquito itself does not bite humans.

[11] W. E. Bradshaw & C. M. Holzapfel (2001).

[12] C. D. Thomas *et al.* (2001).

[13] It is possible that this evolutionary selection process does not provide the full explanation for why morph frequencies have changed in these species. Climatic conditions are known to directly affect whether cricket nymphs will mature to become long- or short-winged adults, so it is plausible that higher temperatures have caused more nymphs to develop longer wings, producing the observed swing in morph frequencies. It is not well understood which factor is most important in this case, but, regardless, the cricket example demonstrates that dispersal ability can change and that natural selection provides a mechanism by which this can be achieved.

[14] These additional examples of changes in dispersal ability are described in C. D. Thomas *et al.* (2001).

[15] S. M. Rovito *et al.* (2009).

[16] G. Moreno (1989).

[17] J. P. Gibbs & N. E. Karraker (2006).

[18] Darwin devoted the first chapter of *The Origin of Species* (1859) to discussing variation under domestication.

[19] D. K. Skelly *et al.* (2005).

[20] C. Parmesan (2006).

[21] A. Baird & J. A. Maynard (2008), O. Hoegh-Guldberg *et al.* (2008a).

Chapter 8

[1] J. P. Harmon *et al.* (2009).

[2] U.S. National Assessment Synthesis Team (2000).

[3] K. B. Suttle *et al.* (2007).

[4] I. Sterling *et al.* (1999).

[5] See M. G. Dyck *et al.* (2008) and references therein.

[6] R. F. Rockwell & L. J. Gormezano (2009).

[7] Robert Rockwell, pers. comm.

[8] A. R. Ives & B. J. Cardinale (2004).

[9] R. H. Drent & J. Prop (2008).

[10] E. O. Wilson (2001).

[11] P. C. de Ruiter *et al.* (2005).

[12] The case for "whole-ecosystem conservation" has been put forward by A. R. Ives & B. J. Cardinale (2004).

[13] M. Scheffer *et al.* (2001).

[14] M. Gladwell (2002).

[15] J. H. Brown *et al.* (1997).

Chapter 9

[1] Independent reviews of the email scandal and IPCC data errors are provided by M. Russell et al. (2010) and Netherlands Environmental Assessment Agency (2010), respectively. For discussion of recent slippage in the public's trust of scientists, see Anon (2010a) and J. Tollefson (2010).

[2] C. D. Thomas *et al.* (2004).

[3] My assessment of newspaper articles follows a study led by Richard Ladle that focused on reports in the UK print media: R. J. Ladle *et al.* (2004, 2005). My findings are in general agreement with the analysis by Ladle

and his colleagues. However, Ladle and his colleagues found a higher proportion of articles reporting the "one million species" claim (21 out of 29 articles, compared to 24 out of 39 in my analysis) and a higher proportion of articles reporting that extinctions would occur *by* 2050 (13 out of 29, compared to 9 out of 39 in my analysis). The 39 newspaper reports that I looked at were obtained through a database search performed by Mai Qaraman, research services librarian at the American Museum of Natural History. The search was undertaken in August 2009 using the Dow Jones Factiva and the Gale InfoTrac Custom Newspapers databases, and was a full-text search for articles including the words "Thomas" and "extinct" and the phrase "climate change" between the dates of January 1 and 31, 2004. These searches returned 39 articles from different newspaper outlets (newspapers with different names and distributions). Text is often similar or identical among reports, since it was taken directly from the newswires (most often the Associated Press) or was shared between newspapers that have common ownership.

The newspaper sources were, **from the U.S.:** the *Akron Beacon Journal,* the *Baton Rouge Advocate,* the *Boston Globe,* the *Capital Times & Wisconsin State Journal,* the *Chicago Sun-Times, Duluth News Tribune, Fort Wayne Journal Gazette,* the *Kansas City Star, Newsday* (NY), the *New York Times,* the *Pittsburgh Post-Gazette,* the *San Francisco Chronicle,* the *Seattle Times,* the *State* (Columbia, SC), *St. Louis Post-Dispatch,* the *Tallahassee Democrat, Tulsa World, USA Today,* the *Washington Post.* **From the UK:** the *Citizen, Daily Mail, Evening Standard,* the *Guardian,* the *Independent,* the *Scotsman, Western Daily Press, Yorkshire Post.* **From Australia:** the *Age* (Melbourne), the *Canberra Times,* the *Gold Coast Bulletin, Herald Sun* (Melbourne), *Townsville Bulletin.* **From Canada:** *Edmonton Journal,* the *Hamilton Spectator, National Post,* the *Toronto Star, Winnipeg Free Press.* **From India:** the *Hindu.* **International:** *International Herald Tribune.*

[4] This paragraph is based on R. J. Ladle *et al.* (2004, 2005). The comment by Margot Wallström is from M. Wallström (2004). The quotes from WWF

(UK) are from a letter sent to members, January 12, 2004.

[5] R. J. Ladle *et al.* (2004, 2005).

[6] R. J. Ladle *et al.* (2004, 2005).

[7] Q. Schiermeier (2004b). Note that surveys in the U.S. and UK have found contrasting results to the one from Germany, suggesting that the movie did raise public concern over climate change—see A. Balmford *et al.* (2004), and A. A. Leiserowitz (2004)—yet the study from Germany that I quote suffices to illustrate the potential risks of exaggeration.

[8] L. Hannah & B. Phillips (2004), C. D. Thomas (2004).

[9] For a similar argument, see R. J. Ladle *et al.* (2004, 2005) and M. Hulme (2006).

[10] A. Revkin (2008).

[11] P. A. Harrison *et al.* (2001).

[12] Published November 14, 2001.

[13] Published January 26, 2002.

[14] R. A. Pielke Jr. (2007).

[15] A complete webcast of John Holdren's nomination hearing, held February 12, 2009, is available via the website of the U.S. Senate Committee on Commerce, Science, and Transportation: http://commerce.senate.gov.

[16] R. J. Ladle *et al.* (2004, 2005).

[17] Al Gore's speech can be viewed at: www.aaas.org/meetings/2009/program/lectures/gore.shtml. See also D. Grimm (2009).

[18] Quote taken from Gore's 2009 AAAS speech; see www.aaas.org/meetings/2009/program/lectures/gore.shtml.

[19] R. Dallek (2009).

[20] R. A. Pielke Jr. (2007).

[21] R. A. Pielke Jr. (2007).

Chapter 10

[1] T. E. Curry *et al.* (2007).

[2] R. A. Kerr (2009).

[3] IPCC (2007b).

[4] B. W. Brook *et al.* (2008).

[5] J. M. J. Travis (2003).

[6] D. Scott *et al.* (2002), D. Scott (2005).

[7] IUCN (International Union for Conservation of Nature) (2005).

[8] C. J. Raxworthy *et al.* (2008). See Chapter 2.

[9] D. Hole *et al.* (2009).

[10] S. G. Willis *et al.* (2009).

[11] One example of a well-intentioned relocation going awry comes from Newfoundland, where red squirrels were introduced in 1963 in the hope that they would provide much-needed prey for declining pine marten populations. Unfortunately, an unforeseen consequence of the introduction was that the squirrels competed with local birds for food, resulting in near extinction of the Newfoundland red crossbill. See M. Schwartz (2005).

[12] O. Hoegh-Guldberg *et al.* (2008b).

[13] Bioclimate envelope models were used by Steve Willis and his team; S. G. Willis *et al.* (2009).

[14] J. J. Tewksbury *et al.* (2002).

[15] P. F. Donald & A. D. Evans (2006).

[16] C. C. Chester (2006).

[17] To find out more about the Great Eastern Ranges Initiative in Australia, see www.greateasternranges.org.au. For further information about the Mesoamerican Biological Corridor project, see A. López and A. Jiménez (2007).

[18] A. Bowkett (2008).

[19] E. Marris (2008).

[20] See the 2009 Technology, Entertainment, Design (TED) talk by Jonathan Drori: www.ted.com/talks/jonathan_drori_why_we_re_storing_billions_ of_seeds.html.

[21] M. S. Swaminathan (2009).

[22] See the 2009 TED talk by Jonathan Drori: www.ted.com/talks/jonathan_drori_why_we_re_storing_billions_of_seeds.html.

[23] J. R. Mawdsley *et al.* (2009).

[24] D. Normile (2009).

[25] O. Hoegh-Guldberg *et al.* (2007).

[26] I recommend the following books for discussion about ways to mitigate climate change: G. Walker & D. King (2008); and E. Mathez (2009), see, in particular, Chapter 10.

[27] Y. Malhi *et al.* (2008).

[28] J. Tollefson (2009).

[29] N. Stern (2007).

[30] A. James *et al.* (2001). The $8 billion figure that I present has been updated for 2010 to account for inflation.

[31] A. Balmford *et al.* (2002). The $53 billion figure that I present has been updated for 2010; see B. M. Hillmann and J. Barkmann (2009).

[32] B. M. Hillmann and J. Barkmann (2009).

[33] Anon (2010b), R. Costanza *et al.* (2010).

[34] P. Sukhdev (2009).

[35] The World Bank (2009).

[36] J. Tollefson (2009).

[37] J. Tollefson (2009).

[38] See section 6 of the Copenhagen Accord, available at http://unfccc.int/resource/docs/2009/cop15/eng/l07.pdf.

[39] P. Sukhdev (2009).

[40] Pages 94–95 in G. Walker & D. King (2008).

[41] This point has been made by Achim Steiner, executive director of the United Nations Environment Programme (UNEP), in a presentation to the Second Diversitas Open Science Conference, Cape Town, South Africa, October 14, 2009. (Transcript available at: www.diversitas-international.org/docs/DIV-OSC2_Speech_A%20Steiner.pdf).

REFERENCES

——. (2010) A question of trust. *Nature*, 466:7.

——. (2010) A full accounting. *Nature*, 465:985–986.

Ahas, R. (1999) Long-term phyto-, ornitho- and ichthyophenological time-series analyses in Estonia. *International Journal of Biometeorology*, 42:119–123.

Baird, A. & J. A. Maynard. (2008) Coral adaptation in the face of climate change. *Science*, 320:315–316.

Balmford, A., A. Bruner, P. Cooper, *et al.* (2002) Economic reasons for conserving wild nature. *Science*, 297:950–953.

Balmford, A., A. Manica, L. Airey, *et al.* (2004) Hollywood, climate change, and the public. *Science*, 305:1713.

Beaubien, E. G. & H. J. Freeland. (2000) Spring phenology trends in Alberta, Canada: links to ocean temperature. *International Journal of Biometeorology*, 44:53–59.

Beaugrand, G., K. M. Brander, J. A. Lindley, *et al.* (2003) Plankton effect on cod recruitment in the North Sea. *Nature*, 426:661–664.

Beaumont, L. J., I. A. W. McAllan & L. Hughes. (2006) A matter of timing: changes in the first data of arrival and last date of departure of Australian migratory birds. *Global Change Biology*, 12:1339–1354.

Berkelmans, R., G. De'ath, S. Kininmonth, *et al.* (2004) A comparison of the 1998 and 2002 coral bleaching events on the Great Barrier Reef: spatial correlation, patterns, and predictions. *Coral Reefs*, 23:74–83.

Blaustein, A. R. & A. Dobson. (2006) A message from the frogs. *Nature*, 439:143–144.

Both, C. & M. Visser. (2001) Adjustment to climate change is constrained by arrival date in a long-distance migrant bird. *Nature*, 411:296–298.

Both, C., S. Bouwhiuis, C. M. Lessells, *et al.* (2006) Climate change and population declines in a long-distance migratory bird. *Nature*, 441:81–83.

Bowkett, A. (2008) Recent captive-breeding proposals and the return of the ark concept to global species conservation. *Conservation Biology*, 23:773–776.

Bradley, N. L., A. C. Leopold, J. Ross, *et al.* (1999) Phenology changes reflect climate change in Wisconsin. *Proceedings of the National Academy of Sciences U.S.A.*, 96:9701–9704.

Bradshaw, W. E. & C. M. Holzapfel. (2001) Genetic shift in photoperiodic response correlated with global warming. *Proceedings of the National Academy of Sciences U.S.A.*, 98:14509–14511.

Brook, B. W., N. S. Sodhi & C. J. A. Bradshaw. (2008) Synergies among extinction drivers under climate change. *Trends in Ecology and Evolution*, 23:453–460.

Brown, J. H., T. J. Valone & C. G. Curtin. (1997) Reorganization of an arid ecosystem in response to recent climate change. *Proceedings of the National Academy of Sciences U.S.A.*, 94:9729–9733.

Carpenter, K. E., M. Abrar, G. Aeby, *et al.* (2008) One-third of reef-building corals face elevated extinction risk from climate change and local impacts. *Science*, 321:560–563.

Catovsky, S. & F. A. Bazzaz. (1999) Elevated CO_2 influences the response of two birch species to soil moisture: implications for forest community structure. *Global Change Biology*, 5:507–518.

Chapin III, F. S., E. S. Zavaleta, V. T. Eviner, *et al.* (2000) Consequences of changing biodiversity. *Nature*, 405:234–242.

Chen, X., B. Hu & R. Yu. (2005) Spatial and temporal variation of phenological growing season and climate change impacts in temperate eastern China. *Global Change Biology*, 11:1118–1130.

Chester, C. C. (2006) *Conservation Across Borders: Biodiversity in an Interdependent World*. Island Press.

Chmielewski, F-M, A. Müller & E. Bruns. (2004) Climate change and trends in phenology of fruit trees and field crops in Germany, 1961–2000. *Agricultural and Forest Meteorology,* 12:69–78.

Colwell, R., G. Brehm, C. L. Cardelús, *et al.* (2008) Global warming, elevational range shifts, and lowland biotic attrition in the Wet Tropics. *Science,* 322:258–261.

Costanza, R., D. Batker, J. Day *et al.* (2010) The perfect spill: solutions for averting the next Deepwater Horizon. *Solutions,* June 16. (available at www.thesolutionsjournal.com/node/629)

Curry, T. E., S. Ansolabehere & H. Herzog. (2007) *A survey of public attitudes towards climate change and climate change mitigation technologies in the United States: analyses of 2006 results.* Laboratory for Energy and the Environment, Massachusetts Institute of Technology; Cambridge, MA.

Dallek, R. (2009) All the President's scholarly men. *Nature,* 458:572–573.

Darwin, C. (1859) *On the Origin of Species by Means of Natural Selection.* John Murray Publishers.

Davies, Z. G., R. J. Wilson, S. Coles, *et al.* (2006) Changing habitat associations of a thermally constrained species, the silver-spotted skipper butterfly, in response to climate warming. *Journal of Animal Ecology,* 75:247–256.

Davis, M. B., R. G. Shaw & J. R. Etterson. (2005) Evolutionary responses to climate change. *Ecology,* 86:1704–1714.

Davis, M. B. & R. G. Shaw. (2001) Range shifts and adaptive responses to Quaternary climate change. *Science,* 292:673–679.

Derocher, A. E., N. J. Lunn & I. Stirling. (2004) Polar bears in a warming climate. *Integrative and Comparative Biology,* 44:163–176.

de Ruiter, P. C., V. Wolters, J. C. Moore, *et al.* (2005) Food web ecology: playing Jenga and beyond. *Science,* 309:68–71.

Donald, P. F. & A. D. Evans. (2006) Habitat connectivity and matrix restoration: the wider implications of agri-environment schemes. *Journal of Applied Ecology,* 43:209–218.

REFERENCES

Drent, R. H. & J. Prop. (2008) Barnacle goose *Branta leucopsis* surveys on Nordenskiöldkysten, west Spitzbergen 1975–2007: breeding in relation to carrying capacity and predator impact. *Circumpolar Studies*, 4:59–83.

Dyck, M. G., W. Soon, R. K. Baydack. *et al.* (2008) Reply to response to Dyck *et al.* (2007) on polar bears and climate change in western Hudson Bay by Sterling *et al.* (2008). *Ecological Complexity*, 5:289–302.

Edwards, M. & A. J. Richardson. (2004) Impact of climate change on marine pelagic phenology and tropic mismatch. *Nature*, 430:881–884.

Elith, J., C. H. Graham, R. P. Anderson, *et al.* (2006) Novel methods improve prediction of species' distributions from occurrence data. *Ecography*, 29:129–151.

Fitter, A. H. & R. S. R. Fitter. (2002) Rapid changes in flowering time in British plants. *Science*, 296:1689–1691.

Foden, W., G. F. Midgley, G. Hughes, *et al.* (2007) A changing climate is eroding the geographic range of the Namib Desert tree *Aloe* through population declines and dispersal lags. *Diversity and Distributions*, 13:645–653.

Gazeau, F., C. Quiblier, J. M. Jansen, *et al.* (2007) Impact of elevated CO_2 on shellfish calcification. *Geophysical Research Letters*, 34:L07603.

Gibbs, J. P. & N. E. Karraker. (2006) Effects of warming conditions in eastern North American forests on red-backed salamander morphology. *Conservation Biology*, 20:913–917.

Gladwell, M. (2002) *The Tipping Point*. Back Bay Books.

Goodman, S. & J. P. Benstead, editors. (2003) *The Natural History of Madagascar*. University of Chicago Press.

Grimm, D. (2009) Al Gore to Scientists: "We Need You." *Science*, 323:998.

Hall-Spencer, J. M., R. Rodolfo-Metalpa, S. Martin, *et al.* (2008) Volcanic carbon dioxide vents show ecosystem effects of ocean acidification. *Nature*, 454:96–99.

Hannah, L. & B. Phillips. (2004) Extinction risk coverage is worth inaccuracies. *Nature*, 430:141.

Hannah, L., G. Midgley, G. Hughes, *et al.* (2005) The view from the Cape: Extinction risk, protected areas, and climate change. *Bioscience*, 55:231–242.

Harmon, J. P., N. A. Moran & A. R. Ives. (2009) Species response to environmental change: impacts of food web interactions and evolution. *Science*, 323:1347–1350.

Harrison, P. A., P. M. Berry & T. P. Dawson, editors. (2001) *Climate change and nature conservation in Britain and Ireland: Modeling natural resource responses to climate change (the MONARCH project)*. UKCIP Technical Report, Oxford.

Hawkins, S. J. (2006) Impacts of climate change on intertidal species. In P. J. Buckley, *et al.*, editors, *Marine Climate Change Impacts Annual Report Card 2006*. Online Summary Reports, MCCIP, Lowestoft. (available at www.mccip.org.uk)

Hawkins, S. J., A. J. Southward & M. J. Genner. (2003) Detection of environmental change in a marine ecosystem—evidence from the western English Channel. *The Science of the Total Environment*, 310:245–256.

Hillmann, B. M. & J. Barkmann. (2009) Conservation: a small price for long-term economic well-being. *Nature*, 461:37.

Hitch, A. T. & P. L. Leberg. (2007) Breeding distributions of North American bird species moving north as a result of climate change. *Conservation Biology*, 21:534–539.

Hoegh-Guldberg, O. (1999) Climate change, coral bleaching and the future of the world's coral reefs. *Marine & Freshwater Research*, 50:839–866.

Hoegh-Guldberg, O., P. J. Mumby, A. J. Hooten, *et al.* (2007) Coral reefs under rapid climate change and ocean acidification. *Science*, 318:1737–1742.

Hoegh-Guldberg, O., P. J. Mumby, A. J. Hooten, *et al.* (2008a) Coral adaptation in the face of climate change: Response to Baird & Maynard. *Science*, 320:315–316.

Hoegh-Guldberg, O., L. Hughes, S. McIntyre, *et al.* (2008b) Assisted colonization and rapid climate change. *Science*, 321:345–346.

Hole, D., S. G. Willis, D. J. Pain, *et al.* (2009) Projected impacts of climate change on a continent-wide protected area network. *Ecology Letters*, 12:420–431.

Hulme, M. (2006) Chaotic world of climate truth. *BBC Online News*: http://news.bbc.co.uk/go/pr/fr/-/2/hi/science/nature/6115644.stm, 4 November.

Huntley, B. (2005) Northern Temperate Responses. Pages 109–124 in T. E. Lovejoy & L. Hannah, editors, *Climate Change and Biodiversity*. Yale University Press.

Huntley, B. & H. J. B. Birks. (1983) *An Atlas of Past and Present Pollen Maps for Europe: 0-13000 B.P.* Cambridge University Press.

IPCC. (2007a) *Climate Change 2007: Synthesis Report. Contribution of Working Groups I, II and III to the Fourth Assessment Report of the Intergovernmental Panel on Climate Change* (Core Writing Team, Pachauri, R.K & Reisinger, A., editors). IPCC, Geneva, Switzerland, 104 pp. (Available at www.ipcc.ch)

IPCC. (2007b) *Climate Change 2007: The Physical Science Basis. Contribution of Working Group I to the Fourth Assessment Report of the Intergovernmental Panel on Climate Change* (Solomon, S., D. Qin, M. Manning *et al.*, editors). Cambridge University Press, 996 pp. (available at www.ipcc.ch)

IUCN (International Union for Conservation of Nature). (2005) Benefits beyond boundaries. Proceedings of the Fifth IUCN World Parks Congress, IUCN, Gland, Switzerland.

Ives, A. R. & B. J. Cardinale. (2004) Food-web interactions govern the resistance of communities after non-random extinctions. *Nature*, 429:74–177.

Jackson, S. T. & C. Weng. (1999) Late Quaternary extinction of a tree species in eastern North America. *Proceedings of the National Academy of Sciences U.S.A.*, 96:13847–13852.

James, A., K. J. Gaston & A. Balmford. (2001) Can we afford to conserve biodiversity? *Bioscience*, 51:43–52.

Jones, G. V. & R. E. Davis. (2000) Climate influences on grapevine phenology, grape composition, and wine production and quality for Bordeaux, France. *American Journal of Enology and Viticulture*, 51:249–261.

Kerr, R. A. (2009) Amid worrisome signs of warming, "climate fatigue" sets in. *Science*, 326:926–927.

La Marca, E., K. R. Lipps, S. Lötters, *et al.* (2005) Catastrophic population declines and extinctions in Neotropical harlequin frogs (Bufonidae: *Atelopus*). *Biotropica*, 37:190–201.

Ladle, R. J., P. Jepson, M. B. Araújo, *et al.* (2004) Dangers of crying wolf over risk of extinctions. *Nature*, 428:799.

Ladle, R. J., P. Jepson & R. J. Whittaker. (2005) Scientists and the media: the struggle for legitimacy in climate change and conservation science. *Interdisciplinary Science Reviews*, 30:231–240.

Leiserowitz, A. A. (2004) Before and after *The Day After Tomorrow*. *Environment*, 46:22–37.

Lerman, J. C. & E. M. Cigliano. (1971) New carbon-14 evidence for six hundred years old *Canna compacta* seed. *Nature*, 232:568–570.

Levitus, S., J. Antonov, T. Boyer. (2005) Warming of the world ocean, 1955–2003. *Geophysical Research Letters*, 32:L02604.

Lipps, K. R., J. Diffendorfe, J. R. Mendelson III, *et al.* (2008) Riding the wave: Reconciling the roles of disease and climate change in amphibian declines. *PLoS Biology*, 6:441–454.

López, A. & A. Jiménez. (2007) *The Mesoamerican Biological Corridor as a Mechanism for Transborder Environmental Cooperation*. Report of the Regional Consultation, 4–5 July 2006, Mexico City, UNEP.

Magurran, A. E. (2005) Biological diversity. *Current Biology*, 15:R116–R118.

Malcolm, J., C. Liu, R. P. Neilson, *et al.* (2006) Global warming and extinctions of endemic species from biodiversity hotspots. *Conservation Biology*, 20:538–548.

Malhi, Y., J. T. Roberts, R. A. Betts, *et al.* (2008) Climate change, deforestation, and the fate of the Amazon. *Science*, 319:169–172.

Marris, E. (2008) Bagged and boxed: it's a frog's life. *Nature*, 452:394–395.

Mathez, E. (2009) *Climate Change: The Science of Global Warming and Our Energy Future.* Columbia University Press.

Mawdsley, J. R., R. O'Malley & D. S. Ojima. (2009) A review of climate-change adaptation strategies for wildlife management and biodiversity conservation. *Conservation Biology*, 23:1080–1089.

May, R. M. (2010) Tropical arthropod species: more or less? *Science*, 329:41–42.

Midgley, G. F., L. Hannah, D. Millar, *et al.* (2002) Assessing the vulnerability of species richness to anthropogenic climate change in a biodiversity hotspot. *Global Ecology and Biogeography,* 11:445–451.

Moreno, G. (1989) Behavioral and physiological differentiation between the color morphs of the salamander, *Plethodon cinereus. Journal of Herpetology*, 23:335–341.

Musil, C., U. Schmiedel & G. F. Midgley. (2005) Lethal effects of experimental warming approximating a future climate change scenario on southern African quartz-field succulents: a pilot study. *New Phytologist*, 165:539–547.

Myers, N., R. A. Mittermeier, C. G. Mittermeier, *et al.* (2000) Biodiversity hotspots for conservation priorities. *Nature*, 403:853–858.

Nadkarni, N. M. & N. T. Wheelwright, editors. (2000) *Monteverde: Ecology and Conservation of a Tropical Cloud Forest.* Oxford University Press.

Netherlands Environmental Assessment Agency. (2010) *Assessing an IPCC assessment: An analysis of statements on projected regional impacts in the 2007 report.* The Hague/Bilthoven.

Normile, D. (2009) Bringing coral reefs back from the living dead. *Science*, 325:559–561.

Parmesan, C. & G. Yohe. (2003) A globally coherent fingerprint of climate change impacts across natural systems. *Nature*, 421:37–42.

Parmesan, C. (2006) Ecological and evolutionary responses to recent climate change. *Annual Review of Ecology, Evolution, and Systematics*, 37:637–669.

Pearson, R. G. & T. P. Dawson. (2003) Predicting the impacts of climate change on the distribution of species: are bioclimate envelope models useful? *Global Ecology and Biogeography*, 12:361–371.

Perry, A. L., P. J. Low, J. R. Ellis, *et al.* (2005) Climate change and distribution shifts in marine fishes. *Science*, 308:1912–1915.

Peterson, A. T., M. A. Ortega-Huerta, J. Bartley, *et al.* (2002) Future projections for Mexican faunas under global climate change scenarios. *Nature*, 416:626–629.

Pielke Jr., R. A. (2007) *The Honest Broker.* Cambridge University Press.

Pounds, J. A. (2001) Climate and amphibian declines. *Nature,* 410:639–640.

Pounds, J. A. & M. L. Crump. (1994) Amphibian declines and climate disturbance: The case of the golden toad and the harlequin frog. *Conservation Biology*, 8:72–85.

Pounds, J. A. & L. A. Coloma. (2008) Beware the lone killer. *Nature Reports Climate Change*, 2:57–59.

Pounds, J. A., M. P. L. Fogden & J. H. Campbell. (1999) Biological response to climate change on a tropical mountain. *Nature*, 398:611–615.

Pounds, J. A., M. R. Bustamante, L. A. Coloma, *et al.* (2006) Widespread amphibian extinctions from epidemic disease driven by global warming. *Nature*, 439:161–167.

Prentice, I. C., A. Bondeau, W. Cramer, *et al.* (2007) Dynamic global vegetation modeling: quantifying terrestrial ecosystem responses to large-scale environmental change. Pages 175–192 in J. G. Canadell *et al.*, editors, *Terrestrial Ecosystems in a Changing World.* Springer.

Raup, D. (1991) *Extinction: Bad genes or bad luck?* W.W. Norton & Company.

Raxworthy, C. J., R. G. Pearson, N. Rabibisoa, *et al.* (2008) Extinction vulnerability of tropical montane endemism from warming and upslope displacement: a preliminary appraisal for the highest massif in Madagascar. *Global Change Biology*, 14:1–18.

Reaser, J., *et al.* (2000) Coral bleaching and global climate change: scientific findings and policy recommendations. *Conservation Biology*, 14:1500–1511.

Reusswig, F. (2005) The international impact of *The Day After Tomorrow*. *Environment*, 47:41–42.

Revkin, A. (2008) Vanishing frogs, climate, and the front page. The *New York Times* Dot Earth blog: http://dotearth.blogs.nytimes.com, 24 March.

Revkin, A. (2009) Hacked e-mail is new fodder for climate dispute. *New York Times*, 20 November.

Richardson, A. J. (2008) In hot water: zooplankton and climate change. *ICES Journal of Marine Science*, 65:279–295.

Rockwell, R. F. & L. J. Gormezano. (2009) The early bear gets the goose: climate change, polar bears and lesser snow geese in western Hudson Bay. *Polar Biology*, 32:539–547.

Rohr, J. R. & T. R. Raffel. (2010) Linking global climate and temperature variability to widespread amphibian declines putatively caused by disease. *Proceedings of the National Academy of Sciences U.S.A.*, 107:8269–8274.

Ron, S., W. E. Duellman, L. A. Coloma, *et al.* (2003) Population decline of the Jambato toad *Atelopus ignescens* (Anura: Bufonidae) in the Andes of Ecuador. *Journal of Herpetology*, 37:116–126.

Root, T. L., J. T. Price, K. R. Hall, *et al.* (2003) Fingerprints of global warming on wild animals and plants. *Nature*, 421:57–60.

Rosenthal, E. (2010) U.N. panel's glacier warning is criticized as exaggerated. *New York Times*, 18 January.

Rosenzweig, C., D. Karoly, M. Vicarelli, *et al.* (2008) Attributing physical and biological impacts to anthropogenic climate change. *Nature*, 453:353–357.

Rovito, S. M., G. Parra-Olea, C. R. Vásquez-Almazán, *et al.* (2009) Dramatic declines in neotropical salamander populations are an important part of the global amphibian crisis. *Proceedings of the National Academy of Sciences U.S.A.*, 106:3231–3236.

Russell, M., G. Boulton, P. Clarke *et al.* (2010) *The Independent Climate Change Emails Review*, 7 July. (available at www.cce-review.org)

Sallon, S., E. Solowey, Y. Cohen, *et al.* (2008) Germination, genetics, and growth of an ancient date seed. *Science*, 320:1464.

Savage, J. M. (2000) The Discovery of the Golden Toad. Pages 171–172 in N. M. Nadkarni & N. T. Wheelwright, editors, *Monteverde: Ecology and Conservation of a Tropical Cloud Forest*. Oxford University Press.

Scheffer, M., S. Carpenter, J. A. Foley, *et al.* (2001) Catastrophic shifts in ecosystems. *Nature*, 413:591–596.

Schiermeier, Q. (2004a) Climate findings let fisherman off the hook. *Nature*, 428:4.

Schiermeier, Q. (2004b) Disaster movie highlights transatlantic divide. *Nature*, 431:4.

Schwartz, M. (2005) Conservationists should not move *Torreya taxifolia*. *Wild Earth*, winter 2005:73–79.

Scott, D. (2005) Integrating climate change into Canada's national park system. Pages 342–345 in T. E. Lovejoy & L. Hannah, editors, *Climate Change and Biodiversity*. Yale University Press.

Scott, D., J. R. Malcolm & C. Lemieux. (2002) Climate change and modeled biome representation in Canada's national park system: implications for system planning and park mandates. *Global Ecology and Biogeography*, 11:475–484.

Sheppard, C. R. C. (1999) Coral decline and weather patterns over 20 years in the Chagos Archipelago, Central Indian Ocean. *Ambio*, 28:472–478.

Skelly, D. K., L. N. Joseph, H. P. Possingham, *et al.* (2005) Evolutionary responses to climate change. *Conservation Biology*, 21:1353–1355.

Southward, A. J. & D. J. Crisp. (1956) Fluctuations in the distribution and abundance of intertidal barnacles. *Journal of the Marine Biological Association UK*, 35:211–229.

Southward, A. J., S. J. Hawkins & M. T. Burrows. (1995) Seventy years' observations of changes in distribution and abundance of zooplankton and intertidal organisms in the Western English Channel in relation to rising sea temperature. *Journal of Thermal Biology*, 20:127–155.

Southward, A. J., O. Langmead, N. J. Hardman-Mountford, *et al.* (2005) Long-term oceanographic and ecological research in the Western English Channel. *Advances in Marine Biology*, 47:1–105.

Spalding, M. D., *et al.* (2001) *World Atlas of Coral Reefs.* University of California Press.

Sparks, T. & P. Carey. (1995) The responses of species to climate over two centuries: an analysis of the Marsham phenological record, 1736–1947. *Journal of Ecology*, 83:321–329.

Sterling, E. J., M. M. Hurley & L. D. Minh. (2006) *Vietnam: A Natural History.* Yale University Press.

Sterling, I., N. J. Lunn & J. Iacozza. (1999) Long-term trends in the population ecology of polar bears in western Hudson Bay in relation to climatic change. *Arctic*, 52:294–306.

Stern, N. (2007) *The Economics of Climate Change: The Stern Review.* Cambridge University Press.

Stuart, S. M., J. S. Chanson, N. A. Cox, *et al.* (2004) Status and trends of amphibian declines and extinctions worldwide. *Science*, 306:1783–1786.

Sukhdev, P. (2009) Costing the Earth. *Nature*, 462:277.

Suttle, K. B., M. A. Thomsen & M. E. Power. (2007) Species interactions reverse grassland responses to climate change. *Science*, 315:640–642.

Swaminathan, M. S. (2009) Gene banks for a warming planet. *Science*, 325:517.

Taylor, S. G. (2008) Climate warming causes phenological shift in Pink Salmon, *Oncorhynchus gorbuscha*, behavior at Auke Creek, Alaska. *Global Change Biology*, 14:229–235.

Tewksbury, J. J., D. J. Levey, N. M. Haddad, *et al.* (2002) Corridors affect plants, animals, and their interactions in fragmented landscapes. *Proceedings of the National Academy of Sciences U.S.A.*, 99: 12923–12926.

Thomas, C. D. (2004) Publish, publicise, and be damned. *Bulletin of the British Ecological Society*, 35:14–16.

Thomas, C. D. & J. J. Lennon. (1999) Birds extend their ranges northwards. *Nature*, 399:213.

Thomas, C. D., E. J. Bodsworth, R. J. Wilson, *et al.* (2001) Ecological and evolutionary processes at expanding range margins. *Nature*, 411:577–581.

Thomas, C. D., A. Cameron, R. E. Green, *et al.* (2004) Extinction risk from climate change. *Nature*, 427:145–148.

Thomas, C. D., A. M. A. Franco & J. K. Hill. (2006) Range retractions and extinction in the face of climate warming. *Trends in Ecology and Evolution*, 21:415–416.

Thuiller, W., S. Lavorel, M. B. Araújo, *et al.* (2005) Climate change threats to plant diversity in Europe. *Proceedings of the National Academy of Sciences U.S.A.*, 102:8245–8250.

Tollefson, J. (2009) Paying to save the rainforests. *Nature*, 460:936–937.

Tollefson, J. (2010) An erosion of trust. *Nature*, 466:24–26.

Travis, J. M. J. (2003) Climate change and habitat destruction: a deadly anthropogenic cocktail. *Proceedings of the Royal Society B*, 270:467–473.

U.S. National Academy of Sciences. (2000) *El Niño and La Niña: Tracing the Dance of Ocean and Atmosphere.* (available at www7.nationalacademies.org/opus/elnino.html)

U.S. National Assessment Synthesis Team. (2000) *Climate Change Impacts on the United States: The Potential Consequences of Climate Variability and Change.* U.S. Global Change Research Program, Washington, D.C.

Vavrek, M. C., J. B. McGraw & C. C. Bennington. (1991) Ecological genetic variation in seed banks. III. Phenotypic and genetic differences between young and old seed populations of *Carex bigelowii. Ecology*, 79:645–662.

Visser, M. & C. Both. (2005) Shifts in phenology due to global climate change: the need for a yardstick. *Proceedings of the Royal Society B*, 272:2561–2569.

Visser, M. & L. J. M. Holleman. (2001) Warmer springs disrupt the synchrony of oak and winter moth phenology. *Proceedings of the Royal Society of London B*, 268:289–294.

REFERENCES

Visser, M., L. J. M. Holleman & P. Gienapp. (2006) Shifts in caterpillar biomass phenology due to climate change and its impact on the breeding biology of an insectivorous bird. *Oecologia*, 147:164–172.

Voyles, J., S. Young, L. Berger, *et al.* (2009) Pathogenesis of chytridiomycosis, a cause of catastrophic amphibian declines. *Science*, 326:582–585.

Walker, G. & D. King. (2008) *The Hot Topic: What we can do about global warming.* Houghton Mifflin Harcourt.

Wallström, M. (2004) In the *Guardian* (London) Society Pages, January 21, p. 12.

Weber, E. (2001) Current and potential ranges of three exotic Goldenrods (*Solidago*) in Europe. *Conservation Biology*, 15:122–128.

Weber, E. & B. Schmid. (1998) Latitudinal population differentiation in two species of *Solidago* (Asteraceae) introduced into Europe. *American Journal of Botany*, 85:1110–1121.

Whitfield, S. M., K. E. Bell, T. Philippi, *et al.* (2007) Amphibian and reptile declines over 35 years at La Selva, Costa Rica. *Proceedings of the National Academy of Sciences U.S.A.*, 104:8352–8356.

Wilkinson, C. R. & D. Souter, editors. (2008) *Status of Caribbean coral reefs after bleaching and hurricanes in 2005*. Global Coral Reef Monitoring Network and Reef and Rainforest Research Centre, Townsville, Australia.

Wilkinson, C. R. (2000) World-wide coral reef bleaching and mortality during 1998: a global climate change warning for the new millennium? Pages 43–57 in C. Sheppard, editor, *Seas at the Millennium: an environmental evaluation (volume III)*. Elsevier.

Wilkinson, C. R., editor. (2004) *Status of Coral Reefs of the World 2004*. Australian Institute of Marine Science. (available at www.gcrmn.org)

Wilkinson, C. R, O. Linden, H. Cesar, *et al.* (1999) Ecological and socioeconomic impacts of 1998 coral mortality in the Indian Ocean: an ENSO impact and a warning of future change? *Ambio*, 28:188–196.

Williams, S. E., E. E. Bolitho & S. Fox. (2003) Climate change in Australian tropical rainforests: an impending environmental catastrophe. *Proceedings of the Royal Society of London B*, 270:1887–1892.

Willis, S. G., J. K. Hill, C. D. Thomas, *et al.* (2009) Assisted colonization in a changing climate: a test-study using two U.K. butterflies. *Conservation Letters*, 2:45–51.

Wilson, E. O. (2001) *The Diversity of Life*. Penguin Books.

Woodward, F. I. (1990) The impact of low temperatures in controlling the geographical distribution of plants. *Philosophical Transactions of the Royal Society of London B*, 326:585–593.

The World Bank. (2009) *State and Trends of the Carbon Market 2009*. Washington, D.C.

Yom-Tov, Y. & S. Yom-Tov. (2004) Climate change and body size in two species of Japanese rodents. *Biological Journal of the Linnean Society*, 82:263–267.

FURTHER READING

Barnosky, A. D. (2009) *Heatstroke: Nature in an Age of Global Warming*. Island Press.

Colbert, E. (2006) *Field Notes from a Catastrophe: Man, Nature, and Climate Change*. Bloomsbury.

Flannery, T. (2005) *The Weather Makers: How Man Is Changing the Climate and What It Means for Life on Earth*. Grove Press.

Hannah, L. (2010) *Climate Change Biology*. Academic Press.

Lovejoy, T. E. & L. Hannah, editors. (2005) *Climate Change and Biodiversity*. Yale University Press.

Lynas, M. (2008) *Six Degrees: Our Future on a Hotter Planet*. National Geographic.

.

CONVERSIONS

When converting a fixed temperature from °F to °C: subtract 32, then multiply by 5, then divide by 9. For example, 68°F = 68 − 32 = 36 x 5 = 180 / 9 = 20°C.

When converting a change in temperature, 1.8°F = 1°C. For example, a temperature increase of 3.6°F is equivalent to an increase of 2°C.

1 millimeter = 0.039 inch
1 centimeter = 0.39 inch
1 meter = 3.28 feet
1 kilometer = 0.62 mile
1 square kilometer = 0.39 square mile, 100 hectares, 247 acres

ACKNOWLEDGMENTS

The science described in this book is the result of the hard work and brilliance of hundreds of researchers, only a fraction of whom are named in the book. I am fortunate to have met or collaborated with many of these scientists, and the ideas, data, and stories that I review here are much more theirs than they are mine. I am especially grateful to Pam Berry and Terry Dawson, who first sparked my interest in the subject, and to the many colleagues who took the time to comment on drafts of sections of the book: Miguel Araújo, Andrew Balmford, William Bradshaw, Barry Brook, Robert Colwell, Wendy Foden, James Gibbs, Lee Hannah, Jason Harmon, Thomas Hickler, Christina Holzapfel, Brian Huntley, Anthony Ives, Richard Ladle, Paul Leburg, Jay Malcolm, Luc te Marvelde, Edmond Mathez, Town Peterson, Roger Pielke, Jr., Alan Pounds, Chris Raxworthy, Anthony Richardson, Robert Rockwell, Cynthia Rosenzweig, Charles Sheppard, Tim Sparks, Eleanor Sterling (who read a draft of the entire book), Melanie

Stiassny, Blake Suttle, Marcel Visser, Ewald Weber, Steve Willis, and Ian Woodward. Suggestions and corrections provided by these readers greatly improved the final manuscript, though, of course, any errors that may remain are my responsibility alone.

My colleagues at the American Museum of Natural History have been instrumental in the book's making. Mai Qaraman and others at the museum's library provided invaluable help. Chris Raxworthy and Eleanor Sterling kindly supported the project, Edmond Mathez engaged in useful discussions, and Niles Eldredge, Michael Novacek, Ian Tattersall, and Neil deGrasse Tyson provided useful advice on getting published. Betina Cochran, Arti Finn, Molly Leff, Alex Navissi, and Caitlin Roxby ensured the project ran smoothly.

I am also immensely grateful to the team at Sterling Publishing, in particular Joelle Herr, for her careful editing, and Pamela Horn.

Finally, my wife, Silvina Vilas, provided extremely valuable comments on a draft of the entire book. Most importantly, she shared in the sacrifice of many evenings and weekends that was necessary in taking on this project. In that regard, this book is very much a joint effort. Thank you.

INDEX

INDEX

Index

ABOUT THE AUTHOR

Richard Pearson is a scientist at the American Museum of Natural History (AMNH), where he is affiliated with the Center for Biodiversity and Conservation, and the Department of Herpetology. His PhD research at Oxford University was on the impacts of climate change on biodiversity, and he has published widely on this subject in leading peer-reviewed scientific journals, including *Nature*. His research at AMNH has been funded by grants from NASA and the National Science Foundation. Completing his doctorate in 2004, Richard moved from Britain to the United States in 2005, where he lives with his wife in New York City.